The Analysis of Drugs of Abuse

ELLIS HORWOOD SERIES IN FORENSIC SCIENCE

Series editor: Dr JAMES ROBERTSON, Head of Forensic Services Division, Australian Federal Police, Canberra, Australia

The Analysis of Drugs of Abuse: an Instruction Manual

M. D. Cole
B. Caddy
Forensic Science Unit, University of Strathclyde

ELLIS HORWOOD
New York   London   Toronto   Sydney   Tokyo   Singapore

First published 1995 by
Ellis Horwood Limited
Campus 400, Maylands Avenue
Hemel Hempstead
Hertfordshire, HP2 7EZ
A division of
Simon & Schuster International Group

Printed and bound in Great Britain by
Bookcraft (Bath) Ltd.

Library of Congress Cataloging-in-Publication Data

Cole, M. D. (Michael D.) .
    The Analysis of drugs of abuse : an instruction manual / M.D.
Cole, B. Caddy.
        p.   cm. -- (Ellis Horwood series in forensic science)
    Includes bibliographical references.
    ISBN 0-13-035098-2
    1. Drugs--Analysis. 2. Chemistry, Forensic. I. Caddy, Brian.
II. Title. III. Series.
RS 189.C583 1995
614' .1--dc20   94-22816
                                                          CIP

British Library Cataloguing in Publication Data

A catalogue record for this book is available
from the British Library

ISBN 0-13-035098-2

---

1 2 3 4 5   99 98 97 96 95

Table of contents

Chapter 3 Presumptive tests for drugs of abuse

Chapter 6 High performance liquid chromatography of drugs of abuse

Chapter 7 Ultra violet spectroscopy of drugs of abuse

Chapter 9   Gas chromatography-mass spectroscopy of drugs of abuse

PREFACE

This book has been developed as part of a training course for instructing students in the analysis of drugs of abuse. This book is not a "recipe" book, rather the emphasis is to instruct those with little or no knowledge of drug analysis and to offer to those with more experience some more challenging aspects of analysis. All the methods described have been worked through by the authors over the past three years and students have had little difficulty in reproducing the analytical processes described. Apart from the bench work the volume is designed to stimulate the student, through a series of questions and answers, into thinking for him/herself. The answers are provided at the end of the book. For the drug chemist to ask the right question can be as important as the analysis. Finally, the student, through this book is stimulated to think of any analyses in the context of the legal process. We hope that this volume will have done something to help in the fight against the ever increasing problem of drug abuse.

M D Cole
B Caddy
Glasgow, 1994

ACKNOWLEDGEMENTS

Many people have contributed to the preparation of this teaching manual, in many ways. However, special thanks must be afforded to the many students and colleagues who have made suggestions and contributions to the course. The Indianapolis-Marion County Forensic Services agency, and Dave Willoughby in particular, are thanked for their help in the preparation of photographs of the drug samples. Frank Drumm, Patricia Flanigan and Ann Coxon have all provided excellent technical support, without which many of the experiments would not be possible. Finally, for the preparation of the manuscript, Veronica Crate must receive special mention.

1

The drug chemist and forensic science

1.1 Introduction

Forensic Science may be defined as the application of scientific principles to the legal process. By this means the forensic scientist helps the courts to reach a decision as to the guilt or innocence of an accused person or persons. Forensic scientists should always consider themselves witnesses for the courts, and not specifically as witnesses for the prosecution or defence. The area of drugs of abuse is only one of a series of subjects embraced by the discipline of forensic science and it is important for the drug chemist to have a full appreciation of his or her role as a forensic scientist.

What is specialist knowledge and how can the courts be assured that the person putting forward the information has the depth of knowledge and expertise to give the courts true guidance for its deliberations? If this is not the case then such "experts" may totally mislead the court with disastrous consequences for Criminal Justice systems and for any accused person or persons. The answer must be in accreditation and certification. The former relates to the analytical processes employed to try and solve the forensic problem, while the latter relates to the competence of the scientist, in our case, the drug chemist, to carry out the analytical procedures and to interpret the results of the analyses in the context of the case.

Most witnesses in any legal process are only able to provide evidence of fact to the Courts, that is to say, they will relay to the court what they saw or what they said or was said to them. Only an expert is permitted to present opinion evidence. This is a very privileged position which must not be abused.

What is an "expert" and who identifies that person as being one? We should think of an expert as someone who is capable of giving expert testimony to a court of law because of knowledge in his or her possession which is not available to the general populace. The question then arises as to what is specialist knowledge and how can the courts be assured that the person putting forward the information has the depth of knowledge and expertise to give the courts true guidance for its deliberations? If this is not the case then such "experts" may totally mislead the court with disastrous consequences for Criminal Justice systems and for any accused person or persons. The answer must be in accreditation and certification. The former relates to the analytical processes employed to try and solve the forensic problem, while the latter relates to the competence of the scientist, in our case, the drug chemist, to carry out the analytical procedures and to interpret the results of the analyses in the context of the case.

1.2 Investigating the scene

Scene investigations for the drug chemist differs little in principle from those of other scene investigations. It is important that only those persons who are really necessary at that scene are, in fact, present. These would normally be a senior police officer and perhaps a junior support, a fingerprint officer, a photographer and a drug chemist. With appropriate training civil crime scenes officers can fulfil most of these roles. In some parts of the world larger teams may be necessary which could include many armed police officers, dogs and their handlers and explosive experts who should have bomb disposal knowledge.

Dogs are particularly useful where large areas are to be searched (airport storage depots), but they soon tire. There are a number of "sniffer" devices available, which operate on several different principles but usually require drug vapours to be sucked into a chamber where they are detected. Typical systems are based on infra-red absorbance or gas chromatographic mobility. Such devices do not usually constitute evidence of drug identification for court purposes. Where the scene is an outside one, such as a <u>Cannabis</u> crop (often surrounded by a legally grown plant) then care must be taken that the "field" has not been mined with explosive or booby trapped in any way.

This latter point can also sometimes be a problem with the clandestine laboratory. As a general rule, it is important to make each scene safe before undertaking a full scientific search and assessment. The scene should be disturbed as little as possible, and the search pattern recorded. Preservation of the scene should become second nature to the investigator. This enables fingerprint and footmark evidence to be recovered, especially where they are present as latent prints and marks. Items should be photographed, in place, against a scale. In addition to photography, sketches of the scene and scale drawings using graph paper often serve a useful purpose. This should include the geographical orientation of the scene. Distances should be clearly marked on the scale drawing as well as the relative positions of items. Searches of cavities beneath floor boards, the plumbing system, including water tank lavatory systems and the immediate sewage system, should be conducted, since drugs and/or apparatus for their synthesis may be concealed in these places. Drugs may have also been flushed away to destroy any evidence of their presence.

Other types of evidence may be recovered and subsequently prove important. Examples of this would be hair and fibre evidence and perhaps body fluid samples such as blood. In addition to bulk samples, trace drug samples should be sought. These may be present on flat surfaces used to apportion drug doses or on the surfaces of items employed to synthesize and/or weigh drug samples. After taking the appropriate controls the surface may be swabbed down with a methanol soaked cotton wool swab, and the swab retained for analysis.

The clandestine laboratory may present other problems. If manufacture is in progress then the system must be "switched off" in the order which will prevent any explosion. Any potentially dangerous chemicals such as lithium aluminium hydride (lithium tetrahydroaluminate) should be neutralised as appropriate. All apparatus should be transported to the investigating laboratory for identification of the contents of the reaction vessels. Chemicals found in the clandestine laboratory, including the final synthetic products should be identified. The investigation of these items should help in answering the following questions:

1. Is a controlled drug present in the reaction vessels?

2. Are intermediate products present in the reaction vessels which indicate the synthetic route?

3. Are the chemicals and apparatus present within the laboratory capable of being used for the illicit synthesis of a controlled drug?

4. What other synthetic uses could be attributed to the chemicals and apparatus present which would not be associated with a controlled drug?

The answers to the first three questions should be available once the analyses are complete but answers to the last may not be so easy. However, every attempt should be made to ascertain the possibilities since this is the kind of question for which the courts will require an answer.

Examples of other items of importance which may be recovered from clandestine laboratories and drug distribution centres are chemical recipe books and drawings of chemical formulae, scales, knives (especially for apportioning Cannabis resin), polythene bags, often of the continuous roll variety and electrically heated bar sealers for polythene bags. Physical examination of some of these items before any attempt at chemical analysis is made can often be rewarding. For example, the striation marks on the edge of a block of Cannabis resin may be matched with a particular knife. Similarly, striation marks and interference patterns under polarised light are useful for matching polythene bags, perforation marks produced at the junction of two bags may also produce a match.

As with all scenes, items recovered should be properly packaged using strong paper and polythene bags which are fastened securely preventing both illegal entry and cross contamination. Fresh plant material should not be retained in polythene bags because the sample will rot. It should be collected in paper bags, weighed, dried and reweighed prior to chemical analysis. Importantly each exhibit should bear a label fastened to the item in such a way that it does not become detached.   The label should bear the following information:

1.   Date of collection

2.   Original location of the exhibit.

3.   Type of case to which it relates.

4.   Name of any accused to which the items relates.

5.   Brief description of the item.

6.   Name and signature of person who collected the items.   Room should also be available on the label for the signatures of all those who, in any way, have had possession of that article, no matter how briefly, until it reaches the court for the purposes of the trial. Continuity of the evidence from the scene to the court room is essential.   At some time during the processing of an item its label may also receive one or more coded numbers.   In the more sophisticated laboratory this could be of the bar code variety.

1.3 Record keeping

Full details of what the drug chemist saw and when, at the scene, must be recorded.   In this respect still photographs, but especially video recordings, are most important.   However, this should not preclude the use of a hand written record of events.   Tape recordings can be valuable at a scene but these should only be used in such a way as to comply with any necessary legal requirements - they suffer from the fact that they can be manipulated.   As a rule it is probably better to take more notes (including sketches and drawings) than may be thought absolutely necessary.   Notes taken at a scene may not be as neat and tidy as the scientist would like but they may have to be inspected by the courts at sometime in the legal proceedings.   Importantly, for most legal systems, these notes should be contemporaneous.

Within the laboratory the drug chemist must record from whom each item was received thereby maintaining the continuity of evidence.   He/she will be required to sign any attached label.   A full description of the time received should be recorded and the item's condition on receipt. Especially important is any sign that the item has been tampered with or is open to contamination.

Full experimental details should be recorded. Such detailed analytical information is set out in other parts of this book but it is important here to record some general analytical considerations.

1.4 General analytical considerations

Records should be available, and regularly updated, on the precise methods of analysis being used together, where appropriate, with an assessment of the performance of any instrumentation employed. Instrumental performance should be regularly assessed (and logged) using national and preferably international standards. The reproducibility, accuracy, sensitivity and specificity of any analytical technique must be first established before it is incorporated into any laboratory routine. This refers as much to colour tests as it does to mass spectroscopic techniques. Even when an analytical method has been accepted, the analytical parameters must be checked from time to time and all records referring to the checking retained.

Bench reagents are in regular use within a laboratory but a record of when and how they were prepared and by whom should always be available. Some important forensic cases have had serious ramifications arising from failure to properly record the strengths of reagents.

Having established these analytical criteria it is important that the analyst does not adopt a "black box" mentality. Each analysis still requires the analyst to devote full care and attention to what he/she is trying to achieve. This can be difficult to maintain where many drug analyses are being performed and the results of any tests are recorded on pro-forma. Ticking boxes on such a document still requires the analyst to think.

1.5 Drug regulations, the legal process and report writing

The results of any analyses must not only be intepreted in the context of the case but also against a background of drug regulations. Drug laws in most countries separate offences into the possession of a controlled substance, possession with intent to supply the drug to other persons and trafficking in drugs. The illicit manufacture/cultivation of controlled substances is also forbidden. These activities are treated with increasing seriousness, some countries having a proscribed death sentence for supplying a drug. It is important therefore for the drug chemist to provide information to the courts which will enable the type(s) of offence to be clearly defined.

What is a controlled substance? A controlled substance is a substance listed in the drug act of a particular country for the purposes of controlling its use because of the dangers they hold for society if they are not controlled. For example, for the United Kingdom, the Misuse of Drugs Act 1971 (MDA 1971), and the Misuse of Drugs Act Regulations of 1985 are the Acts mainly responsible for identifying a controlled drug. While most of the commonly abused drugs are specifically listed in these Acts [Table 1.1.] care has been taken to frame the wording of the Act to cover chemical compounds which may not yet be available. For example, in the MDA 1971 reference is made to "ether and ester derivatives" and to "sterioisomers". The advantages of framing an Act in this way are to overcome the problem of "designed drugs". These are drugs which have been illicitly synthesised and for which there are no specific regulations to control their use. The introduction of a modification to an Act is usually such a lengthy process that by the time the "new" drug has been incorporated into legislation it has ceased to be of importance and has probably been superceded by a compound of slightly different chemical structure which is again not specifically covered by legislation.

As well as drugs which are controlled in each individual country's legislation there are drugs whose control has been agreed under international agreement. These drugs are detailed by the United Nations Division of Narcotics [1]. These briefly list compounds under, narcotics - those related to diacetylmorphine; stimulants - the amphetamines; hallucinogens - lysergic acid diethylamide; cocaine and related compounds; barbiturates and benzodiazepines.

In most cases the drug chemist will have prepared a report for the court to which he/she will usually have reference during the course of the trial. It is important that the report conveys the necessary information to the court in a clear and concise manner and without the use of scientific jargon. There are no absolutes as to what is the correct way to write a report but there are certainly incorrect ways. It is most important that the report makes no statement of guilt or innocence to the offence of which the defendant(s) is charged. This is for the court to decide. Great care must be taken not to usurp the role of the court in the decision making process. It is important that the report is a balanced one and that statements are only made which can be substantiated either by experiment or by scientific reports which are in the public domain. For example, a phrase which commonly occurs is "...this is consistent with...". This phrase on its own can be misleading because the facts may also be consistent with some other explanation and these should also be recorded. Better that the word consistent is not used.

In setting out a report there is certain information which has to be presented. Firstly the full name and qualifications of the scientist must be detailed somewhere. The time (date) and from whom items were received and to where they were delivered should also be recorded. It is then usual to detail the exhibits; what they are purported to be, the details of any labels attached to them and a description. Any peculiarities about the appearance should be recorded.

**Table 1.1 : List of drugs in the United Kingdom Misuse of Drugs Act 1971
and the Misuse of Drugs Act Regulations 1985**

List of controlled substances according to Misuse of Drugs Act 1971

Controlled Drugs - Part I

Class A Drugs

1.   The following substances and products, namely:-

Acetorphine
Allylprodine
Alphacetylmethadol
Alphameprodine
Alphamethadol
Alphaprodine
Anileridine

Benzethidine
Benzylmorphine (3-benzylmorphine)
Betacetylmethadol
Betameprodine
Betamethadol
Betaprodine
Bezitramide
Bufotenine

Cannabinol, except where contained
    in cannabis or cannabis resin
Cannabinol derivatives
Clonitazene
Coca leaf
Cocaine

Desomorphine
Dextromoramide
Diamorphine
Diampromide
Diethylthiambutene
Dihydrocodeinone
    O-carboxymethyloxime
Dihydromorphine
Dimenoxadole
Dimepheptanol

Dimethylthiambutene
Dioxaphetyl butyrate
Diphenoxylate
Dipipanone

Ecgonine, and any derivative of
    ecgonine which is convertible
    to ecgonine or to cocaine
Ethylmethylthiambutene
Etonitazene
Etoxeridine

Fentanyl
Furethidine

Hydrocodone
Hydromorphinol
Hydromorphone
Hydroxypethidine

Isomethadone

Ketobemidone

Levomethorphan
Levomoramide
Levophenacylmorphan
Levorphanol
Lysergamide
Lysergide and other N-alkyl
    derivatives of lysergamide

Mescaline
Metazocine

Methadone

Methadyl acetate

Methyldihydromorphine
(6-methyldihydromorphine)

Metopon

Morpheridine

Morphine

Morphine methobromide,
morphine N-oxide and other
pentavalent nitrogen morphine
derivatives

Myrophine

Nicodicodine (6-nicotinoyldi-
hydrocodeine)

Nicomorphine (3,6-dinicotinoyl-
morphine)

Noracymethadol

Norlevorphanol

Normethadone

Normorphine

Norpipanone

Opium, whether raw, prepared
or medicinal

Oxycodone

Oxymorphone

Pethidine

Phenadoxone

Phenampromide

Phenazocine

Piminodine

Piritramide

Poppy-straw and concentrate of
poppy-straw

Proheptazine

Properidine (1-methyl-4-phenyl-
piperidine-4-carboxylic acid
isopropyl ester)

Psilocin

Racemethorphan

Racemoramide

Racemorphan

Thebacon

Thebaine

Trimeperidine

4-Cyano-2-dimethylamino-4,
4-diphenylbutane

4-Cyano-1-methyl-4-phenyl
piperidine

N,N-Diethyltryptamine

N,N-Dimethyltryptamine

2,5-Dimethoxy-α, 4-dimethyl-
phenethylamine

1-Methyl-4-phenylpiperidine-4-
carboxylic acid

2-Methyl-3-morpholino-1,
1-diphenylpropanecarboxylic
acid

4-Phenylpiperidine-4-carboxylic
acid ethyl ester

2.   Any stereoisomeric form of a substance for the time being specific in paragraph 2 above not being dextromethorphan or dextrorphan.

3.   Any ester or ether of a substance for the time being specified in paragraph 1 or 2 above.

4.   Any salt of a substance for the time being specified in any of paragraphs 1 to 3 above.

5.   Any preparation or other product containing a substance or product for the time being specified in any of paragraphs 1 to 4 above.

6.   Any preparation designed for administration by injection which includes a substance or produce for the time being specified in any of the paragraphs 1 to 3 of Part II of this Schedule.

Part II

Class B Drugs

1.   The following substances and products, namely:-

Acetyldihydrocodeine	Methylamphetamine
Amphetamine	Methylphenidate
Cannabis and cannabis resin	Nicocodine
Codeine	Norcodeine
Dexamphetamine	Phenmetrazine
Dihydrocodeine	Pholcodine
Ethylmorphine (3-ethylmorphine)	

2.   Any stereoisomeric form of a substance for the time being specified in paragraph 1 of this Part of this Schedule.

3.   Any salt of a substance for the time being specified in paragraph 1 or 2 of this Part of this Schedule.

4.   Any preparation or other product containing a substance or product for the time being specified in any of paragraphs 1 to 3 of this Part of ths Schedule, not being a preparation falling within paragraph 6 of Part I of this Schedule.

Part III

Class C Drugs

1.   The following substances, namely:-

Benzphetamine	Pemoline
Chlorphentermine	Phendimetrazine
Fencamfamin	Phentermine
Mephentermine	Pipradrol
Methaqualone	Prolintane

2.   Any stereoisomeric form of a substance for the time being specified in paragraph 1 of this Part of this Schedule.

3.   Any salt of a substance for the time being specified in paragraph 1 or 2 of this Part of this Schedule.

4.   Any preparation or other product containing a substance for the time being specified in any of paragraphs 1 to 3 of this Part of this Schedule.

List of controlled substances according to Misuse of Drugs Act Regulations, 1985

SCHEDULE I

Controlled Drugs Subject to the Requirements of Regulations 14, 15, 16, 18, 19, 20, 23, 25 and 26.

1.   The following substances and products, namely:-

(a) Bufotenine
   Cannabinol
   Cannabinol derivatives
   Cannabis and Cannabis resin
   Coca leaf
   Concentrate of poppy-straw
   Eticyclidine
   Lysergamide
   Lysergide and other N-alkyl derivatives of lysergamide
   Mescaline
   Psilocin
   Raw opium
   Rolicyclidine
   Tenocyclidine
   4-Bromo-2, 5-dimethoxy-α-methylphenethylamine
   N,N-Diethyltryptamine
   N,N-Dimethyltryptamine
   2,5-Dimethoxy-α,4-dimethylphenethylamine

(b) any compound (not being a compound for the time being specified in sub-paragraph (a) above) structurally derived from tryptamine or from a ring-hydroxy tryptamine by substitution at the nitrogen atom of the sidechain with one or more alkyl substituents but no other substituents;

(c) any compound (not being methoxyphenamine or a compound for the time being specified in sub-paragraph (a) above) structurally derived from phenethylamine, an N-alkylphenethylamine, α-ethylphenethyl-amine, or an N-alkyl-α-ethylphenethylamine by substitution in the ring to any extent with alkyl, alkoxy, alkylenedroxy or halide substituents, whether or not further substituted in the ring by one or more other univalent substituents.

2.   Any stereiosomeric form of a substance specified in paragraph 1.

3.   Any ester or ether of a substance specified in paragraph 1 or 2.

4.   Any salt of a substance specified in any of paragraphs 1 to 3.

5.   Any preparation or other product containing a substance or product specified in any of paragraphs 1 to 4, not being a preparation specified in Schedule 5.

SCHEDULE 2

Controlled Drugs subject to the Requirements of Regulations 14, 15, 16, 18, 19, 20, 21, 23, 25 and 26

1.   The following substances and products, namely:-

Acetorphine
Alfentanil
Allylprodine
Alphacetylmethadol
Alphameprodine
Alphamethadol
Alphaprodine
Anileridine

Benzethidine
Benzylmorphine (3-benzylmorphine)
Betacetylmethadol
Betacetylmethadol
Betameprodine
Betamethadol
Betaprodine
Bezitramide

Clonitazene
Cocaine

Desomorphine
Dextromoramide
Diamorphine
Diampromide
Diethylthiambutene
Difenoxin
Dihydrocodeinone
   0-carboxymethyloxime
Dihydromorphine
Dimenoxadole
Dimepheptanol
Dimethylthiambutene
Dioxaphetyl butyrate
Diphenoxylate

Dipipanone
Drotebanol

Ecgonine, and any derivative of
  ecgonine which is convertible
  to ecgonine or to cocaine
Ethylmethylthiambutene
Etonitazene
Etorphine
Etoxeridine

Fentanyl
Furethidine

Glutethimide

Hydrocodone
Hydromorphinol
Hydromorphone
Hydroxypethidine

Isomethadone

Ketobemidone

Lefetamine
Levomethorphan
Levomoramide
Levophenacylmorplan
Levorphanol

Medicinal opium
Metazocine
Methadone
Methadyl acetate
Methyldesorphine
Methyldihydromorphine
  (6-methyldihydromorphine)
Metopon
Morpheridine
Morphine
Morphine methobromide, morphine
  N-Oxide and other pentavalent
  nitrogen morphine derivatives

Myrophine

Nicomorphine
Noracymethadol
Norlevorphanol
Normethadone
Normorphine
Norpipanone

Oxycodone
Oxymorphone

Pethidine
Phenadoxone
Phenampromide
Phenazocine
Phencyclidine
Phenomorphan
Phenoperidine
Piminodine
Piritramide
Proheptaxine
Properidine

Racemethorphan
Racemoramide
Racemorphan

Sufetanil

Thebacon
Thebaine
Tilidate
Trimeperidine

4-Cyano-2-dimethylamino-4
  4-diphenylbutane
4-Cyano-1-methyl-4-
  phenylpiperidine
1-Methyl-4-phenylpiperidine-4-
  carboxylic acid
2-Methyl-3-morpholino-1,
  1-diphenylpropanecarboxylic
  acid

4-Phenylpiperidine-4-carboxylic
  acid ethyl ester

2. Any stereoisomeric form of a substance specified in paragraph 1 not being dextromethorphan or dextrorphan.

3. Any ester of ether of a substance specified in paragraph 1 or 2, not being substance specified in paragraph 6.

4. Any salt of a substance specified in any of paragraphs 1 to 3.

5. Any preparations or other product containing a substance or product specified any of paragraphs 1 to 4, not being preparation specified in Schedule 5.

6. The following substances and products, namely:-

Acetyldihydrocodeine	Methylamphetamine
Amphetamine	Methylphenidate
Codeine	Nicocodine
	Nicodicodine
Dextropropoxyphene	(6-nicotinoldihydrocodeine)
Dihydrocodeine	Norcodeine
Ethylmorphine (3-ethylmorphine)	Phenmetrazine
Mecloqualone	Pholcodine
Methaqualone	Propiram

7. Any stereoisomeric form of a substance specified in paragraph 6.

8. Any salt of a substance specified in paragraph 6 or 7.

9. Any preparation or other product containing a substance or product specified any of paragraphs 6 to 8, not being a preparation specified in Schedule 5.

SCHEDULE 3

Controlled Drugs subject to the Requirements of Regulations 14, 15, 16, 18, 22, 23, 24, 25 and 26

1.   The following substances, namely:-

   a) Benzphetamine   Meprobamate
                                        Methylphenobarbitone
      Chlorphentermine   Methyprylone

      Diethylpropion   Pentazocine
                                        Phendimetrazine
      Ethchlorvynol   Phentermine
      Ethinamate   Pipradrol

      Mazindol
      Mephentermine

   b) any 5,5 disubstituted barbituric acid.

2.   Any stereoisomeric form of a substance specified in paragraph 1.

3.   Any salt of a substance specified in paragraph 1 or 2.

4.   Any preparations or other product containing a substance specified in any of paragraphs 1 to 3, not being preparatin specified in Schedule 5.

SCHEDULE 4

Controlled Drugs excepted from the Prohibition on Importation, Exportation and, when in the form of a Medicinal Product, Possession and Subject to the Requirements of Regulations 22, 23, 25 and 26.

1.   The following substances and products, namely:-

   Alprazolam   Fludiazepam
                                        Flunitrazepam
   Bromazepam   Flurazepam

   Camazepam   Halazepam
   Chlordiazepoxide   Haloxazolam
   Clobazam
   Clonazepam   Ketazolam

Clorazepic acid
Clotiazepam
Cloxazolam

Delorazepam
Diazepam

Estazolam
Ethyl loflazepate

Loprazolam
Lormetazepam

Medazepam

Nimetazepam
Nitrazepam
Nordazepam

Oxazepam
Oxazolam

Pinazepam

Prazepam

Temazepam
Tetrazepam
Triazolam

2. Any stereoisomeric form of a substance specified in paragraph 1.

3. Any salt of a substance specified in paragraph 1 or 2.

4. Any preparations or other product containing a substance or product specified in any of paragraphs 1 to 3, not being a preparation specified in Schedule 5.

Views differ as to what should or should not be reported in terms of the analytical data. Certainly in order not to break up the text it is useful to refer to a technique without giving details. For example "...the sample purported to be diacetyl morphine was analysed by high performance liquid chromatography and gas chromatography mass spectrometry...". Full details can be added to the report as an annexation if required.

Finally, the report should contain an opinion of what the analytical results mean in the context of the case. Cases generally fall into one of three categories. The ones concerned with bulk seizures such as may be taken by Customs and Excise personnel; those which relate to a Clandestine laboratory and those in which rather small amounts of drug are concerned. For the first and second types of case there is little doubt that trafficking and manufacture are taking place and opinions can be given on these activities. The last type of case can be more contentious from the analyst's point of view since it can be a fine dividing line between having enough drug for one's own use compared with being in possession of a drug with the intention of supplying to others. The weight of drug becomes important here and the "street" value of the drug at the time of the offence. Common sense judgements and experience can play an important part in reaching an opinion in this type of case. Some legislative systems define the amount above which supply is the criminal charge.

Having provided an opinion all that remains is for the drug analyst to sign the document. It is common practice to sign every page and to mark any page which is not completely covered with print in such a way that nothing can be added to the document. Due to the case work load in some very busy laboratories, reports in the form of pro forma, in which appropriate boxes are completed, has become the norm.

The role of the drug chemist is not yet complete until he/she has presented evidence to a Court of Law. It is advisable that the scientist should have as much laboratory experience as possible before presenting evidence for the first time. Moreover, any experience he/she can obtain from practising lawyers, perhaps using a mock court, should be grasped. Watching experienced forensic scientists presenting evidence is also advantageous. However, the witness box is a lonely place. It should be remembered, whilst you are an expert in relation to drugs, the lawyer is the expert with words. Much has been written about the presentation of evidence at a trial [2] but primarily the rules are to listen to what question is being put, allow time to think of the answer, and have the courage to say when you don't know, no matter how painful that might be. The questioner should be faced. The reply to the judge and jury should be on level terms. A condescending attitude should not be adopted. Answers are always better received if you are properly dressed and adopt a good posture in the witness box. The more experience the drug chemist has at presenting the evidence the better he/she becomes. If there are serious attempts by Counsel to mislead the court on your evidence the Judge is there to help to correct for any misconceptions.

From what has preceded it is hoped that the reader will have developed at least a feeling for the role of the "drug chemist" in an operational forensic science laboratory. However, what has been reported is far from complete and the interested analyst should continue to broaden his or her knowledge base by further readings [3].

The chapters which follow attempt to instil into the reader the analytical techniques which are presently available for establishing the identity of a controlled substance and, where necessary, quantifying the material. A book of this nature cannot record all methods of analyses which have been reported, and there will be many excellent methods which will not be found in this text. What the reader can be assured of, however, is that the methods of analyses recorded here have all been tried and tested by the authors and are known to work.

1.6 References and further reading

[1] Multilingual Dictionary of Narcotic Drugs and Psychotropic Substances under International Control, United Nations, New York, 1983.

[2] Mildred R. H., The Expert Witness, George Godwin, 1982.

[3] Gough, T. A. (ed.), The Analysis of Drugs of Abuse John Wiley & Sons, Chichester, 1991.

[4] Rt. Hon. Sir John Mays, Interim Report on the Maguire Case, House of Commons, HMSO, 12 July 1990.

[5] National Measurement and Accreditation Scheme NIS 46 1 April 1992, NAMAS Executive, National Physical Laboratory, Teddington, Middlesex, TW11 0LW.

[6] "Diplomas in professional subjects", Journal of the Forensic Science Society; 25, 1985, pp. 385-386.

[7] "Diploma in document examination", Journal of the Forensic Science Society; 25, 1985, p. 387.

[8] "Regulations for the Diploma in Firearms Examination", Journal of the Forensic Science Society; 28, 1988, p. 71 and 30, 1990, pp. 383-384.

[9] "Diploma in Crime Scene Investigation", Journal of the Forensic Science Society; 29, 1989, p. 61.

[10] "Regulations for the Diploma in Fire Examination", Journal of the Forensic Science Society; 28, 1988, p. 71.

2

Physical description of drugs of abuse

2.1 Introduction

It is essential that the correct analytical procedure and documentation be followed when dealing with seizures thought to contain drugs of abuse. This maximises the amount of information that can be obtained from the sample, ensures continuity of evidence, and establishes that the whole sample can be accounted for. The first process in the sequence of analysis should be a full physical description of the sample. Recorded data should include the shape, colour, dimensions and weight of the sample; how the items are packaged, and full details of the packaging. The samples should only be analysed if the packaging is correct and undamaged. Information provided on any of the labelling should be recorded. Once the packed samples have been opened, the texture and smell of the contents should be written down. The colour of the sample under white, and under ultraviolet (UV) light should also be recorded.

Packaging comparisons may be made with previous samples, and any microscopical features noted. These should include any imprints, indentations and striations, which may reveal whether a cutting implement or sealing machine (heat seal impressions) has been used on one or more of the same samples. This data may be used to identify seizures with a common origin, an important feature of drugs intelligence. Additionally, where blocks of drug sample have been seized, it should be determined whether or not the pieces fit together.

Seizures which contain crystalline substances should be examined, microscopically, to determine whether the sample is homogenous (more than one crystallization crop may be present), or contains a number of components with different microscopic features.

The following is a general description of the drugs of abuse likely to be encountered on a frequent basis.

2.2 <u>Cannabis</u> <u>sativa</u> L. and products

This drug is now widely abused, and now accounts for over 75% of the drugs cases examined in the United Kingdom. Material from <u>Cannabis</u> may occur in a number of forms, which are best discussed separately.

2.2.1 Whole plant

The whole plant is a strong smelling, glandular plant, growing from 30 cm to 6 m in height. The time from seed to maturity is about 3 months, but the products from the plant may be harvested for up to 2 months after this. The leaves are palmate, with clusters of flowers occuring (Plate [1]). The plant is usually dioecious, that is, male flowers on one plant and female on another, although plants which possess both kinds are known.

The stem is hollow and four cornered, it's surface being covered in anticellular trichomes (hairs), pressed to the stem which curve upward (Plate [2]). The phyllotaxy (leaf arrangement) is alternate and opposite. A thin groove runs along the adaxial (upper) side of the leaf so that the vein forms a U shaped bundle. Axillary buds may be observed.

Each male flower has fine white or greenish hairy sepals (leaf like structures enclosing the flower), about 3.5 mm long, and five pendulous stamens. The flowers hang in loose, multibranched clusters up to 18 cm long. In contrast, the female flowers are borne on racemes, which are borne in pairs. Each flower is surrounded by a bract enwrapping the ovary to form a swollen tube 1.8-2.9 mm long, from which hang two yellow stigmas.

Unicellular (Plate [2]), and warty-walled cystolithic (Plate [3]) trichomes are found on the stem, together with a few glandular trichomes (Plate [4]). On the adaxial surface of the leaves, short, non-glandular broad based trichomes are observed. In contrast, on the abaxial (lower) surface, the hair density is higher and the hairs longer (up to 250 μm). Glandular trichomes can also be seen here.

Material prepared from the flowering tops or leaves is commonly called marijuana although other names can be used. On a weight for weight basis, the dried plant material contains up to 0.7% Δ^9-THC.

2.2.2 Hashish

This is material produced from the leafy aerial tops and flowers of <u>Cannabis</u> <u>sativa</u>. The resin, produced in the glandular trichomes, is scraped from the surface of the plant material, and pressed into blocks. Hashish contains 2-5% by weight of the active constituent, Δ^9-tetrahydrocannabinol (Δ^9-THC), about 10 times that of the leaf material on a weight for weight basis. In this material can be found parts of bracts, bracteoles and trichomes which can be used to identify the plant material, even after smoking (Plate [5]).

Plate 1 Flower clusters of *Cannabis sativa*

Plate 2 Unicellular trichomes of *Cannabis sativa*

Plate 3 Warty walled cystolithic trichomes of *Cannabis sativa*

Plate 4 Glandular trichomes of *Cannabis sativa*

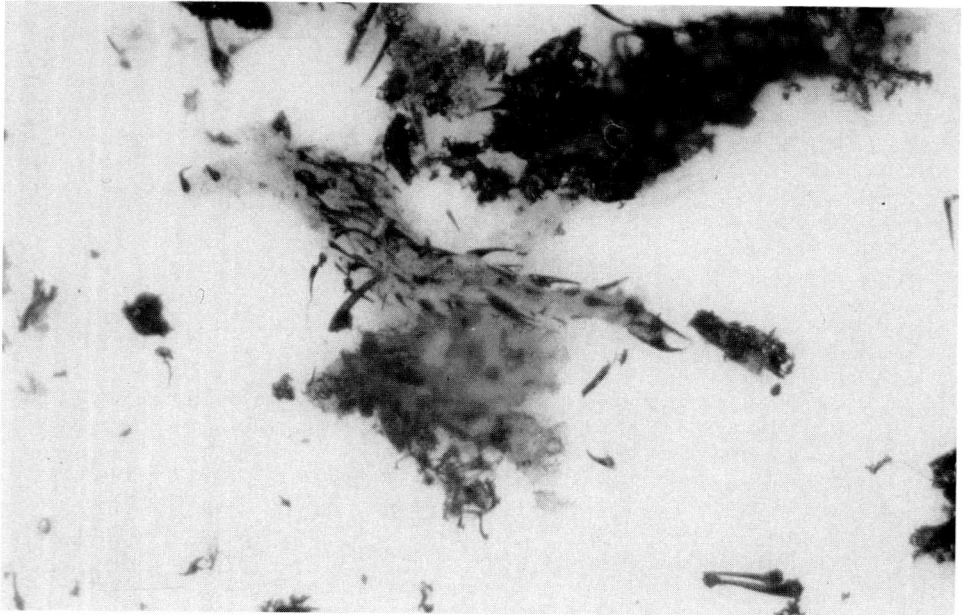

Plate 5 Residues from smoking *Cannabis* material

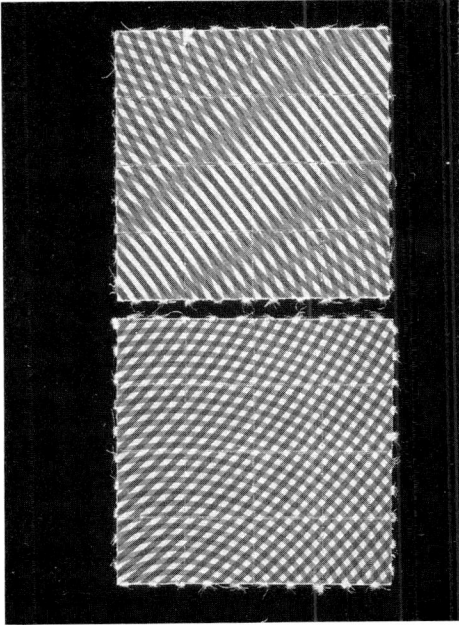

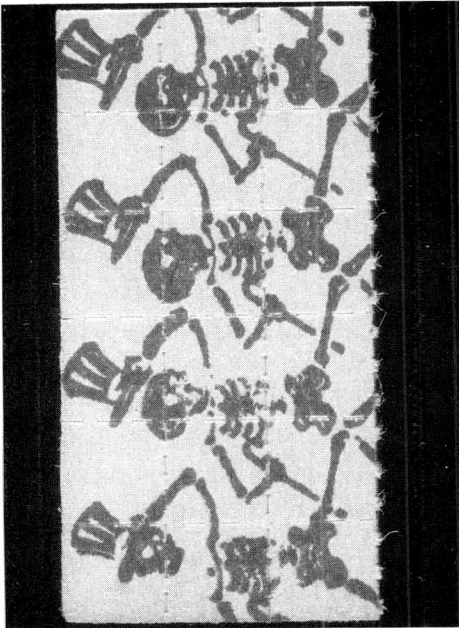

Plate 6 Examples of LSD Blotter acids

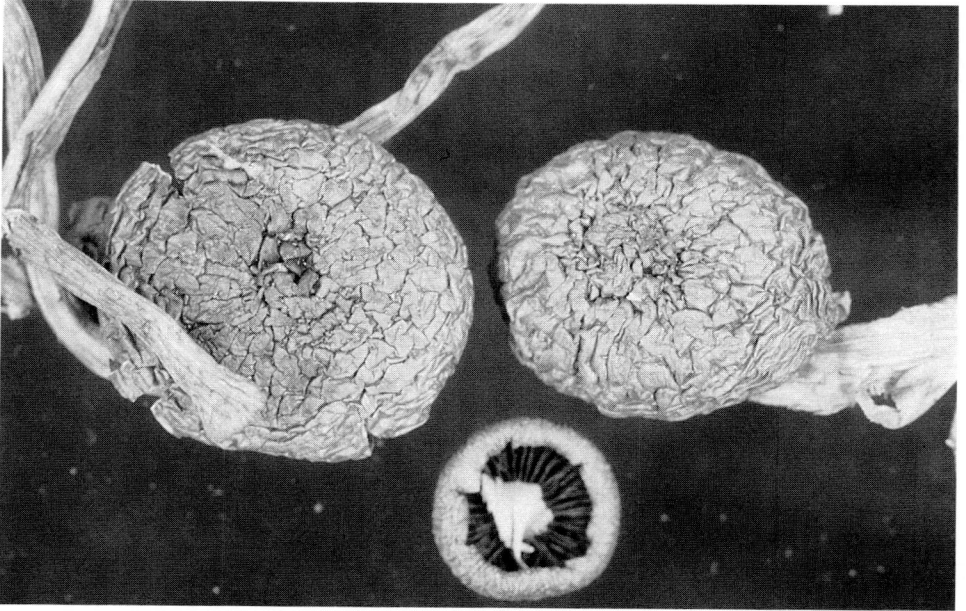

Plate 7 *Psilocybe* mushrooms

Plate 8 Example of 'official' stamp on a block of drug. Note that the stamp is not official at all.

When examining hashish, several characteristics should be recorded. These include whether or not any blocks fit together, such that at one time they could have formed one or more larger blocks. Where necessary striations on the surfaces of blocks of resin should be examined to determine whether they could be associated: for example, whether a common cutting implement such as a knife has been used, leaving characteristic cutting marks which may enable a common origin to be established. If the blocks are broken, the depth and colour of the different layers of the resin should be examined and described. The colours and texture of the resin may be used, in certain cases, to determine the geographical origin of the drugs of abuse. For further details, the reader is referred to the textbook by Gough [1].

When dealing with plant materials and quantifying controlled substances it is important that the plant material is in the dried state and not freshly harvested. The reason for this is that the water content of the plant material will vary according to the physiological state of the plant and the time interval since harvesting, thus affecting any concentration stated using a weight for weight basis.

2.2.3 Hash Oil

Hash oil is manufactured by extraction of the whole plant material with an organic solvent, such as alcohol, ether or benzene. On removal of the solvent, an oil coloured green through brown containing a complex mixture of compounds is obtained. Within this mixture is found Δ^9-THC, the major pharmacologically active drug from Cannabis sativa. On a weight for weight basis, the oil contains in excess of 10% and sometimes up to 50 or 60% Δ^9-THC.

The oil may be used in a number of ways. These include smoking in special pipes and with tobacco.

2.2.4 The active constituents of Cannabis sativa

There is confusion over the nomenclature of the pharmacologically active terpenoids from Cannabis sativa. Two numbering systems are employed, the monoterpene system, and the dibenzofuran system (Figure 2.1.). The latter will be employed throughout this course.

The main pharmacologically active constituent of Cannabis sativa is considered to be Δ^9-tetrahydrocannabinol (Δ^9-THC) (Figure 2.2.). In addition, Δ^9-tetrahydrocannabinolic acid is also found in the resin. This compound is not itself active, but it is readily converted to Δ^9-THC by the heat produced when Cannabis resin is smoked.

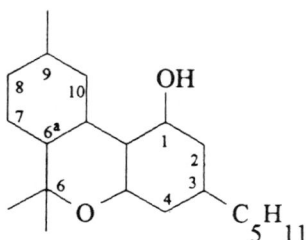

Dibenzofuran system

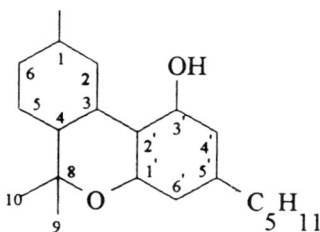

Monoterpenoid system

Fig. 2.1 Numbering systems used for cannabinoids

Δ^9-THC is itself an unstable compound.   There are reports [2] that it will break down completely at 80°C over a period of 7 days and even at ambient temperature 75% will be lost over 10 months.   It is most stable in alcoholic solution but this should be kept away from light.   Care must therefore be taken when storing standard samples purchased from a supplier. Since a piece of resin is not exposed to light or oxygen in its centre, it is from this area, rather than the surface, that samples should be taken for quantification.   The homogeneity of the cannabinoids throughout a block of cannabis resin is much better than might be expected as seen from the data [Table 2.1.] provided by Gough [1].   When compared with the between block variation for the same batch of Cannabis resin the range of concentrations for the individual cannabinoids differs little from the range of values found within a single block.   There are measurable differences between the cannabinoids present in different batches of cannabis resin.

In the plant, Δ^9-THC is considered to be a precurser of cannabinol (CBN) while cannabidiol (CBD) is the precurser of Δ^9-THC.   These and other related cannabinoids also occur as their corresponding carboxylic acids.   It can be seen therefore that any chromatographic analysis for Cannabis, in one of its forms, will produce a complex pattern from which it is usual practice to identify Δ^9-THC,   CBD   and   CBN   and   perhaps   cannabigerol   and cannabichromene.   These will only occur together in the plant and will not be found in Δ^9-THC samples produced by illicit synthesis.   For fuller details on

Cannabidiol

9-tetrahydrocannabinol

8-tetrahydrocannabinol

Cannabinol

9-tetrahydrocannabinolic acid

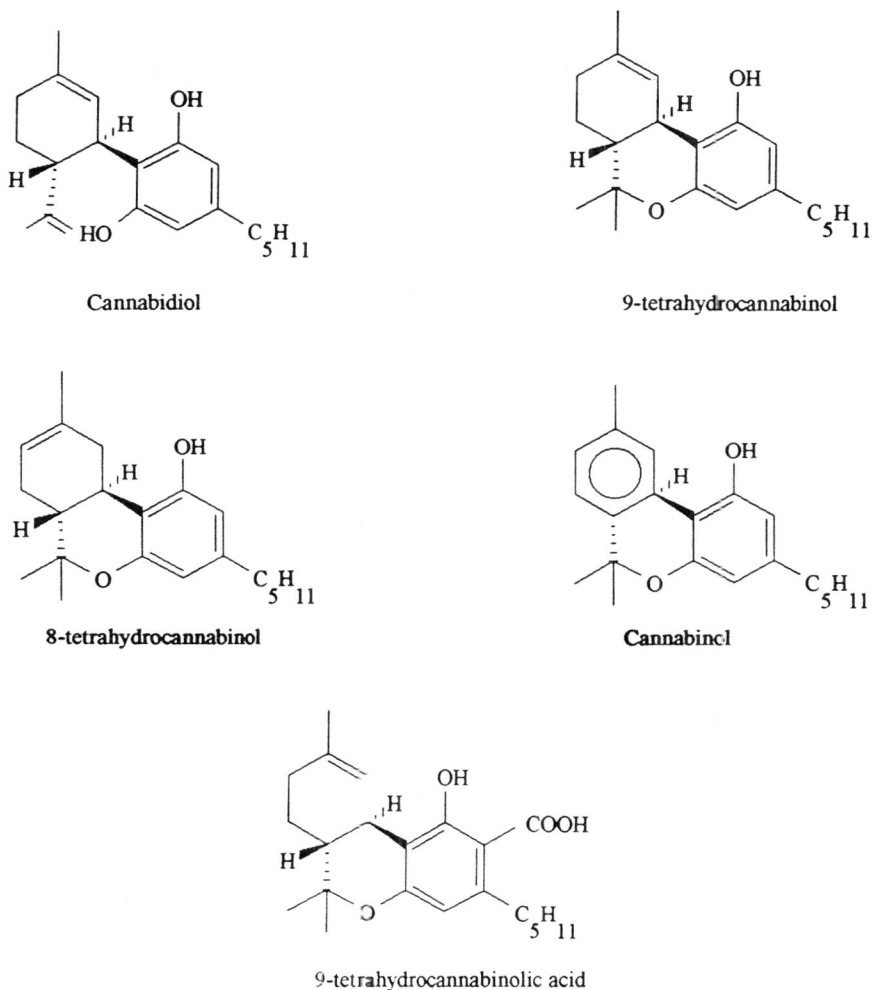

Fig. 2.2 Examples of cannabinoids found in <u>Cannabis</u> <u>sativa</u>

the chemistry of the cannabinoids the reader is referred to the book on <u>Cannabis</u> and Health by Graham [2].

2.3 Lysergic acid diethylamide (LSD)

Lysergic acid diethylamide (N,N-diethyl-d-lysergamide, lysergide) (Figure 2.3.) is one of the most potent hallucinogens known to man. It can be produced by a number of synthetic routes, which include using lysergic acid or ergot alkaloids as starting materials.

$$CO.N(C_2H_5)_2$$

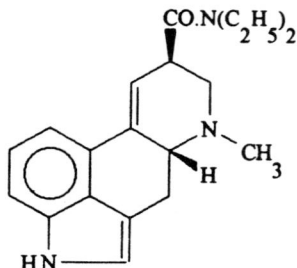

Fig. 2.3 Structure of lysergic acid diethylamide

The drug, in the past, had been applied in solution to a number of inert substrates, including adsorbant paper, sugar cubes, inert material used to fill gelatin capsules or pressed into small tablets commonly called "microdots", or it can be incorporated into a gelatin matrix which is then cut into small pieces. The only common form at the time of writing, however, is that of adsorbant paper impregnated with the drug.

When examining samples thought to contain LSD, extreme caution should be observed since this drug readily penetrates the skin. Protective clothing should be used at all times.

If the sample is composed of sheets of adsorbant paper (examples are shown in plates ([6a - 6d]), the size and number of the dose units, the size and depth of the perforations, the colour and style of the design on the papers should all be measured where possible. Typically, adsorbant paper doses contain 30 - 50 μg LSD per dose unit although higher doses are known. The amount per dosage unit should be determined at the appropriate stage in the analysis of the sample.

If the drug is contained in a powder or microdot, then it should be examined microscopically. A number of different adulterants may be found, which will have different microscopic characteristics, and will be detected in subsequent analyses. These include dextrose, lactose and mannitol, used as diluents; starch, gelatin and ethylcellulose which are used as binders; talc or (magnesium silicate) magnesium stearate, used to lubricate the particulates; maize or potato starch, used to increase volume and usually referred to as fillers; and sometimes, water soluble dyes.

Table 2.1 Intra-slab variation of cannabinoid content of cannabis resin from Pakistan (The analysis of drugs of abuse, Gough, T.A. (ed) 1991, Reprinted with permission of John Wiley and Sons, Ltd).

Compound	Cannabinoid Content Relative to THC = 1.0[a]				
	Interior only	Surface only		slab	
	Resin A (n=6)	Resin A (n=10)		Resin A (n=16)	
	x	x	cv	x	cv
CBD	4.2	4.7	9.0	4.5	9.4
CBG + THV	0.05	0.04	63	0.05	40
CBDA	4.4	5.1	13	4.8	14
CBN	6.8	8.1	13	7.4	14
CBCh	0.3	0.3	-	ND	-
THCA	0.5	0.5	15	0.5	18
	Resin B (n=6)	Resin B (n=10)		Resin B (n=16)	
	x	x	cv	x	cv
CBD	2.5	4.0	13	3.5	24
CBG + THV	0.05	0.04	6.5	0.04	55
CBDA	2.1	3.4	17	2.9	27
CBN	3.1	6.8	20	5.3	42
CBCh	0.2	0.4	50	0.3	53
THCA	0.2	0.3	17	0.2	31
	Resin C (n=6)	Resin C (n=10)		Resin C (n=16)	
	x	x	cv	x	cv
CBD	1.2	1.6	6.1	1.4	12
CBG + THV	0.1	0.1	12	0.1	9.8
CBDA	1.2	1.3	6.3	1.3	8.6
CBN	1.1	1.8	13	1.6	26
CBCh	0.3	0.3	5.2	0.3	9.6
THCA	0.1	0.1	11	0.1	14

[a] n = no of samples taken; x = mean cannabinoid content; cv = coefficient of variation of cannabinoid content; ND = not detected.

2.4 Hallucinogens of fungal origin

Mushrooms are known to contain a number of hallucinogens. Material from such drugs may occur as fresh or dry mushrooms, powders, tablets or capsules. The physiologically active components are nearly all based upon the indole nucleus. Typical examples include psilocybin, and psilocin (Figure 2.4.).

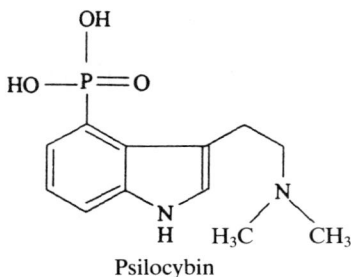

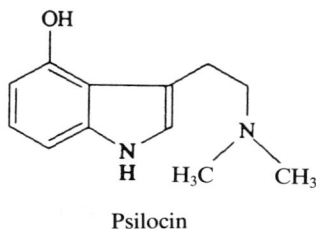

Psilocybin

Psilocin

Fig. 2.4 Examples of hallucinogens from Psilocybe spp.

The main fungal genus which produces psilocybin is Psilocybe. Of the 140 species of Psilocybe, 80 are found to contain psychotropic substances. The two most important species are Psilocybe semilanceata and Psilocybe cubensis. Other psilocybin containing genera include   Panaeolus spp., Conocybe spp., Inocybe spp and Pluteus spp. It is the dephosphorylated compound, psilocin, present in only trace amounts in the fungal material which is the psychoactive principle.

If the seizure contains whole or parts of mushrooms (Plate [7]), the examination of the seizures should include a description of the shape, size, colour and form of the whole fungus, the gills, the spores, the cap and the stipe (stalk). The identification of agaric fungi is very difficult, and should only be performed by experts.   Features which should be noted are described in the United Nations Drug Control Manual ST/NAR/19 [4].

If the material is a powder, the sample should be examined visually and under a microscope, for homogeneity and the presence of adulterants. Any seemingly different components, where possible, should be analysed separately.

2.5 Hallucinogens from cacti

Illicit trafficking in cacti which contain the psychoactive principle, mescaline (Figure 2.5.), is also known.

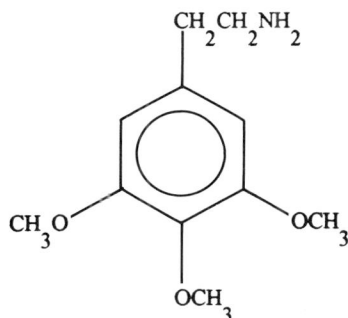

$$CH_2CH_2NH_2$$

$$CH_3O \qquad\qquad OCH_3$$

$$OCH_3$$

Fig. 2.5 Structure of mescaline

This compound occurs in the cacti <u>Trichocereus</u> <u>pachanoi,</u> <u>Trichocereus</u> <u>peruvianus</u> and <u>Lophophera</u> <u>williamsii</u>. Fresh plants of the last of these are blue-grey to greyish green, about 2-7 cm in height and 4-12 cm in diameter, and contains 0.5 - 1.5% mescaline on a drug weight basis, the tops are spineless, in 8 - 10 well defined ribs and furrows. The most common illicit form is produced as dried, disc shaped cuttings taken from the top of the cactus, hence the term "mescal buttons".

2.6 The opiate drugs

2.6.1 Raw opium

Raw opium is produced from the coagulated juice of the poppy, <u>Papaver</u> <u>somniferum</u>, which forms a viscous brown latex after drying for a few hours. Scraped from the seed capsule in this form, the resin is a rich source of opiate alkaloids. The resins from plants from different geographical regions are characterised by different ratios of the opiate alkaloids. For example, Turkish opium shows a low codeine and noscapine level relative to a high morphine level (ratio <0.1) and chinese opium is rich in codeine (3-4%).

2.6.2 Diamorphine

All diacetylmorphine originates from opium, and is prepared by the acetylation of morphine, obtained from opium. A number of extraction techniques for morphine are currently employed. The extraction and acetylation process result in the production of a number of opiate drugs illustrated below (Figure 2.6.). "Heroin" should only be used as a name for the final product mixed with any adulterants which may be present.

No two samples of heroin have an identical appearance since heroin is manufactured in a batch process. In all cases, full details of the seizure should be recorded, and where possible microscopical examination of the sample should be made to ensure it is homogeneous.

Heroin from south eastern Asia is found in a number of forms, many of which are identified by their names. Heroin No.3, also known as smoking heroin, Hong Kong Rocks, or Chinese Heroin, consists of hard granules, 1-10 mm in diameter, which are grey-brown in colour. Small variations in granule size may be observed between each batch. This type of heroin is composed of caffeine and 20-50% diamorphine hydrochloride. In addition, there are coloured pink varieties known as Penang Pink, and Red Chicken. The main impurities of Heroin No 3 of heroin are acetyl codeine and 6-O-monoacetyl morphine. Diluents such as quinine, methaqualone, barbital and caffeine may also be added. This type of heroin may be used for both smoking and injection.

Heroin No. 4, also referred to as injecting heroin, is an off white or white powder, containing about 90% diacetylmorphine hydrochloride with no aggregates.

Heroin from south western Asia originates from the geographical area which lies between the Mediterranean and Pakistan/Afghanistan. Two types predominate [Table 2.2.]. One is a variable beige colour, which occurs as a fine powder, but may contain some soft aggregates. The second type is an off-white colour, and occurs as a fine dry powder and usually contains a higher percentage of diacetylmorphine than the former.

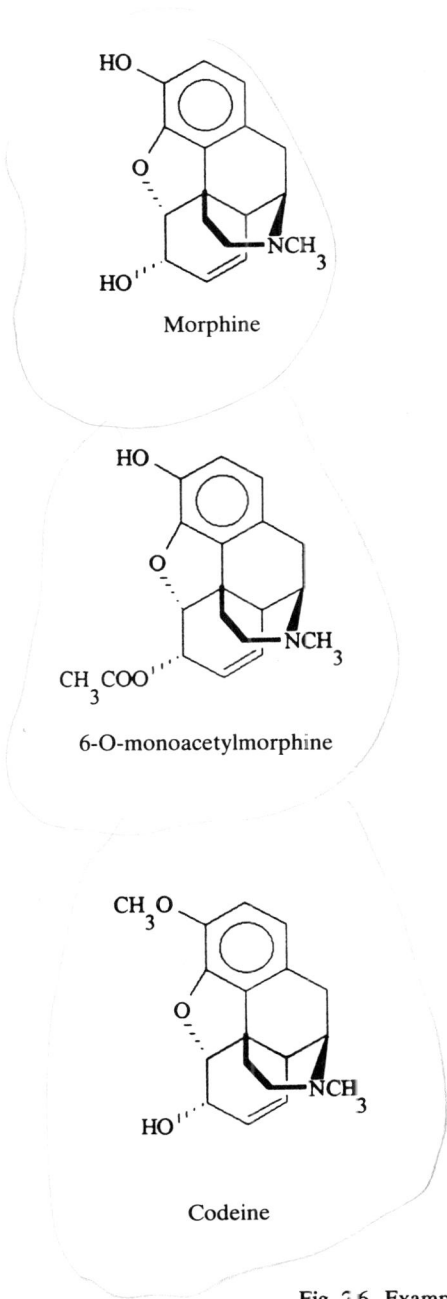

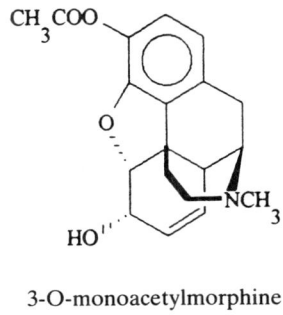

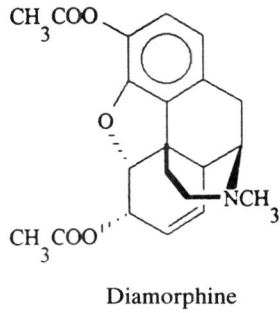

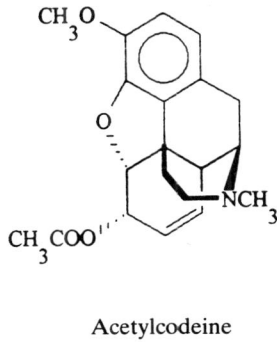

Fig. 2.6 Examples of opiates found in Opium.

Table 2.2 : Typical compositions of South West Asian Heroin.
(DAM = diamorphine, AcCOD = acetylcodeine,
MAM = 6-O-monoacetylmorphine) UNDCP manual ST/NAR/6).

Type	% DAM	% AcCod	% MAM	% Noscapine	% Papaverine
1	60	5	3	10	4
2	80 - 90	3	2	0	0

2.7 Coca leaf and cocaine

The evergreen South American shrub Erythroxylon coca contains a number of tropane alkaloids, of which the most widely known is cocaine (Figure 2.7.). Street samples of cocaine are typically cut with the synthetic local anaesthetics lidocaine and procaine, for example, because they induce a similar effect to cocaine. Caffeine, and/or sugars in addition to other controlled substances may be used as cutting agents.

Cocaine, as the free base, may be extracted from the plant leaves as a viscous oil.   The hydrochloride salt may be isolated by precipitation with hydrochloric acid from an organic solution of the free base.   The salt of cocaine, a fluffy white powder, became known as "snow", for obvious reasons.

The free base, obtained by precipitation with alkali ammonia or bicarbonate from an aqueous solution of the hydrochloride salt is known as "Crack" which has a rather waxy looking, lumpy appearance.   Illicit preparations of the drug may be found wrapped in, for example, foil, small balloon wrappers or plastic bags. The sample should be fully described, and examined for sample homogeneity. Large samples received as blocks may bear official looking stamps (Plate [8]) used in an attempt to make them look legal and authenticated.

Cocaine

Benzoylecgonine

Ecgonine methyl ester

Ecgonine

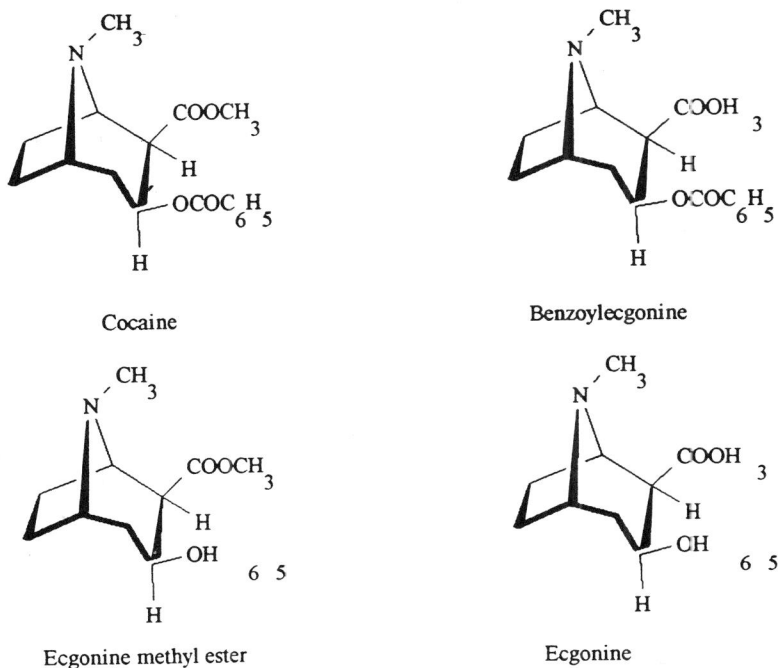

Fig. 2.7 The structure of cocaine and metabolites

2.8 The amphetamines

The amphetamines (Figure 2.8.) are a group of drugs belonging collectively to the phenethylamines. The majority of these drugs were once prescribed for appetite suppression in weight loss programmes.

Others, including amphetamine and methamphetamine are currently used to treat an excessive need for sleep, an illness known as narcolepsy. Each is prescribed officially in tablet or capsule form, but amphetamine and its analogues are often synthesized in clandestine laboratories to produce products which vary in appearance from white powders to grey-brown amorphous looking samples often containing considerable amounts of solvent, precursors, and reaction by-products.

Amphetamines may be synthesized by a number of different routes, and it is these routes of syntheses which will determine the profile of the impurities produced by incomplete and side reactions. Examination of these can be used to link seizures, or trace the origins of the samples. Synthetically prepared analogues of amphetamines are available in many parts of the world. These include the methylenedioxyamphetamines and related compounds.

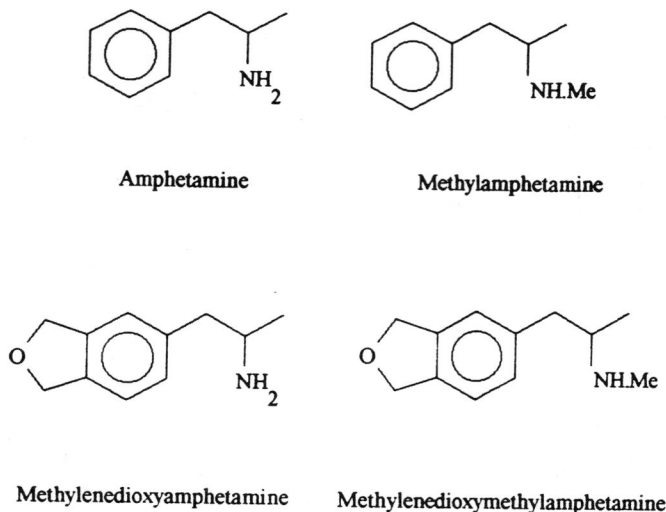

Amphetamine   Methylamphetamine

Methylenedioxyamphetamine   Methylenedioxymethylamphetamine

Fig. 2.8 Structures of amphetamine and derivatives

Pharmaceutically prepared amphetamine is optically pure, that is, it contains only one stereoisomer, whilst the products of illicit synthesis are racemic mixtures, containing equal amounts of the R and S forms. Some drug chemists do however continue to use the old D, L nomenclature to describe the stereochemistry of optically active drugs. Examination of the stereochemistry of the sample using, for example, crystal tests or chiral chromatography, can be used to establish the optical purity of a sample, a factor which can be used to differentiate legal and illicit sources.

2.9 Methaqualone and mecloqualone

Methaqualone and mecloqualone (Figure 2.9.) appear on the illicit drug market either as diversions from the legitimate pharmaceutical trade, or from illicit synthesis. First prepared as a non-addictive, non-barbiturate sleeping drug, methaqualone was used as a legitimately dispensed hypnotic in some European countries.

In the illicit markets, these drugs have a brown or grey appearance and possess the texture of a tacky powder. Depending upon the source the purity ranges from 30% to 70%. Methaqualone is sometimes found as a cutting agent for heroin. The free base or hydrochloride salts of these compounds may be present in capsule, tablet or powder form.

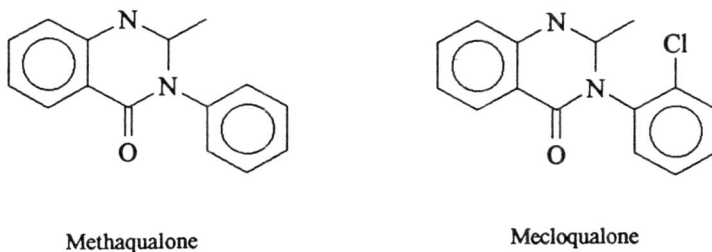

Methaqualone   Mecloqualone

Fig. 2.9 Structure of methaqualone and mecloqualone

The size, dimensions, and marks on the tablets should be noted, and the sample examined microscopically for homogeneity.

2.10 Barbiturates

Under the United Kingdom Misuse of Drugs Act Regulations, 1985, any 5,5-disubstitued barbituric acid is a controlled substance (Figure 2.10.).

They are therapeutically used as sedatives, hypnotics, anaesthetics and anti-convulsants. Virtually all the barbiturates encountered on the illicit market result from the diversion of samples from licit sources. The twelve scheduled barbiturates (UN handbook ST/NAR/18 [5]) occur mainly as capsules and tablets, but some are found as injectable solutions. The compounds may occur as mixtures of barbiturates, or mixed with other drugs such as caffeine, aspirin, ephedrine or codeine.

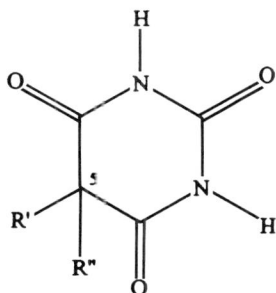

Fig. 2.10 Generalised structure of a barbiturate

If tablet or capsules are encountered, the number, weight, and dimensions of the items should be noted, and the sample examined under the microscope to determine sample homogeneity. If a solution is obtained, a full description of the container and the solution should be recorded.

2.11 Benzodiazepines

These compounds were introduced to replace the barbiturates and methaqualone, and have the general formula shown in Figure 2.11. The benzodiazepines are controlled, by name, under the Misuse of Drugs Act Regulations, 1985.

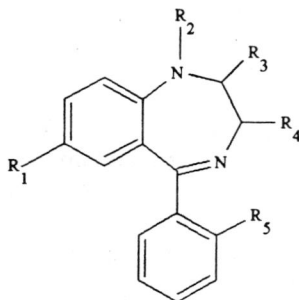

Fig. 2.11   Generalised structure of a benzodiazepine

Such drugs are used as tranquilizers, hypnotics, anticonvulsants and muscle relaxants, and are amongst the most frequently prescribed drugs in western society. All those encountered on the illicit market result from diversion from licit sources. There is no evidence of clandestine manufacture. Those 33 benzodiazepines which currently appear on the control list are produced as tablets and capsules, although some are sold as vials or powders for preparation for injection. Other illicit drug preparations, for example, heroin, may contain benzodiazepine (Temazepam) as a congener.

2.12   Student exercises

2.12.1   Cannabis plant material

You are provided with dried plant material of Cannabis sativa. Examine the leaves, petioles, female flower parts and stems under the low power search microscope.

> **Q.1** Does the material have a characteristic smell?   Is there a difference in trichome density between different plant parts? In which direction do the trichomes lie?   Are glandular trichomes visible?

You are also provided with fresh samples of tobacco, and a sample of hops (Humulus lupulus), the closest relative of Cannabis sativa.

> **Q.2** Do you observe any morphological similarities?   What problems might this cause in sample identification?

2.12.2   Cannabis resin

You are provided with a number of samples of Cannabis resin. Place a small amount of each resin on a cavity microscope slide, and add a few drops of chloral hydrate with a pasteur pipette. With a mounting needle, disperse the suspended plant material. Examine under a low power search microscope.

> **Q.3** Can you recognise any plant parts?   Are some samples easier to disperse than others?

2.12.3   Drugs of fungal origin

You are provided with dried examples of some fungi. Examine each of the samples, and make detailed records of weight, size, shape, colour etc., as you think necessary.

> **Q.4** What are the possible problems encountered in examining dried materials such as these?

2.12.4   Crude opium resin

You are provided with an example of crude opium resin. Write a physical description of the materials.

> **Q.5** What details should you include, and why?   What problems would a large block like this present later on during the chemical analysis stage?

2.12.5   Heroin based powder

You are provided with two samples of opiate drugs. Describe each carefully, and examine under the search microscope.

> **Q.6** Are the samples similar under the microscope - if not, what might this tell you?

2.13   References

[1]   Gough, T. A. (ed.), The Analysis of Drugs of Abuse,   John Wiley & Sons, Chichester, 1991, p.558.

[2]   Graham, J. D. P. (ed.),   Cannabis and Health,   Academic Press, London, 1976, p.481.

[3]   Jackson, B. P. and Snowdon, D. W., Powdered Vegetable Drugs, Stanley Thornes (London) Ltd, 1974, p.207.

[4]   Recommended methods for testing peyote cactus (mescal Buttons)/mescaline and psilocybe mushrooms/psilocybin,   United Nations, New York, 1989.

[5]   Recommended methods for testing barbiturate derivatives under international control, United Nations, New York, 1989.

3

Presumptive tests for drugs of abuse

3.1 Introduction

Having fully described and documented the sample suspected of containing a drug of abuse, the aim becomes to identify the components of the material. Spot tests are the first "presumptive tests" which are used. They have a number of advantages which include ease of use, rapidity of application and cheapness.

Such tests, do, however have a number of disadvantages. They require a correct interpretation of the colour change, which is not usually specific for one compound. This is illustrated in the table below (Table 3.1). This becomes particularly important and relevant when examining street samples, which may not give the same colour reaction as the pure standard due to the presence of adulterants. Finally, they are relatively insensitive when compared to instrumental methods of analyses.

Despite these difficulties, the application of a colour test is a useful first step in the identification of the components of an illicit drug seizure. Whilst not identifying the drug beyond question, the colour reaction may provide important clues as to the identity of the class of compound to which the drug belongs. It has been suggested that the spot tests can be used sequentially to identify the drug of abuse [1], although it should again be emphasized, this is not proof of identify. The degree to which proof of identity of a drug is required depends to some extent upon the legislation of the country in which the forensic scientist is working. However, the application of a large number of colour tests increases the possibility for identifying a particular member of a drug class but this should never be considered absolute proof of identify.

It should also be remembered that when performing the tests, positive and negative controls should also be used. Positive controls, of pure drug standard, verify that the reagents are reacting in the correct way, and the negative controls ensure that "false positives" are not obtained, for example, from solvent residues, improperly washed equipment, and poor laboratory practice.

Table 3.1 Commonly used reagents employed in presumptive tests

Spot Test Reagent	Drugs giving positive results	Colour Reaction
Cobalt iso thiocyanate	Cocaine	Blue
Dille Kopanyi	Barbiturates	Blue
Duquenois Levine	Cannabis products	Blue
Mandelin	Cocaine & surrogates	Orange
	Opiates	Red/Brown/ Olive
Marquis	Amphetamines	Orange/Yellow
	Opiates	Purple/Pink
Ehrlich's Reagent	LSD	Purple/Blue
	Psilocin/Psilocybin	Grey/Purple

The detection limits for the presumptive test lie in the low microgramme range. This is sufficient sensitivity for most street seizures in which relatively large amounts of drug are present. However, when material has been seized which is only thought to have very small traces of material present, for example, syringes, and glassware from clandestine laboratories, such tests should not be used, and more sensitive instrumental methods be used at the outset.

Sampling of the seizure prior to analysis must also be considered. Where a powder sample is obtained, each portion with a different physical appearance should be sampled. When sampling, it is important to ensure that the material being sampled is homogeneous. If a large number of packages (more than ten) are to be sampled, but the total is less than 100, then ten of the packages should be sampled. When the total number of packages is greater than 100, the square root of the number of samples should be sampled.

It should always be remembered that the samples, should, as far as possible, be randomly chosen, and that samples which do not appear similar should not be counted together.

Using this approach, the class of compound to which the drug might belong can be tentatively identified, but it should be remembered that any one reagent may react with more than one class of compound. Such tests will not identify which member of a particular class of compounds is present. To do this, one must use thin layer chromatography, and at least one instrumental technique, to positively identify the drug involved. The tests have recently been summarized in a United Nations Division of Narcotic Drugs Manual [2].

3.2 Tests for common drugs of abuse

3.2.1 Cannabis and products

Two tests are generally used to test for the presence of cannabinoids, namely the Duquenois-Levine test and the Corinth test. They are performed as follows:

3.2.1.1 The Duquenois-Levine test

Three reagents are required:

1. A solution of 2.5ml acetaldehyde and 2g vanillin dissolved in 100ml 95% ethanol

2. Concentrated hydrochloric acid, and

3. Chloroform.

The test is performed by the addition of one volume of the first solution to the test substrate, and shaking the mixture for one minute, then adding one volume of concentrated hydrochloric acid, and three volumes of chloroform. A blue or purple layer will extract into the lower chloroform layer. An alternative method is to prepare a petroleum ether (40-60°C) fraction extract of any material, using 10mg of sample per millilitre of solvent. The test is then applied directly on the extract.

3.2.1.2 The Corinth test

A small amount of the drug sample is placed on a piece of filter paper. A second piece of paper is placed on top of the sample, and a few drops of petroleum ether added (40°C - 60°C fraction). After extracting the drugs from the sample, the upper paper is removed and dried. About 1mg of 1% Corinth V salt in anhydrous sodium sulphate is placed on the filter paper and then a few drops of water are added. The development of a pink colour indicates that cannabinoids may be present. A pink colour should develop in the positive control, but not the negative control. A positive control, using a known sample of Cannabis, should be employed. A negative control, using all of the reagents but no resin, is also necessary.

3.2.2 Lysergic acid diethylamide, psilocybin and psilocin

These hallucinogens can be tested using Ehrlich's reagent (p-Dimethylaminobenzaldehyde, P-DAB).   If an LSD blotter acid seizure has been made, then only one dose unit should be tested.   The reagent is also known as Van Urk's reagent.

The reagent is prepared as 1g para-dimethylaminobenzaldehyde in 10ml methanol, with 10ml concentrated orthophosphoric acid added.   The test is performed by the addition of two drops of reagent to the test substrate, a positive and a negative control.   The colours obtained depend on the drug in question, but include violet (lysergic acid diethylamide), or grey/violet (psilocybin and psilocin).

3.2.3 Opiate drugs

A number of test are available to determine whether or not opiate drugs are present in a sample.   It is, however, difficult to distinguish between the drugs on the basis of this test alone, since many give similar colour reactions in both the Marquis and Mandelin reagents [Table 3.2.].   The tests are performed as shown below.

3.2.3.1   The Marquis test

This reagent is prepared by the addition of 5ml of 40% formaldehyde to 100ml concentrated sulphuric acid.   To perform the test, two drops of reagent are added to the test substrate, a positive and negative control.   A variety of colour reactions will be obtained with opiate drugs, their congeners, and adulterants [Table 3.2.].

3.2.3.2   The Mandelin test

This reagent is prepared as a 1% w/v solution of ammonium metavanadate in 100ml concentrated sulphuric acid.   Two drops of reagent are added to the sample, a positive and negative control.   Again a variety of colours will develop [Table 3.2.].

3.2.4 Cocaine and products

Cocaine can be tested for using a number of reagents, but commonly employed ones are cobalt isothiocyanate reagent, the mandolin reagent, and sodium hydroxide.

Table 3.2 Colour reaction of opiate drugs with Marquis and Mandelin reagents

Drug	Colour on addition of reagent	
	Marquis	Mandelin
Diamorphine	Purple	Red/Brown
6-O-monoacetyl morphine	Pink/Purple	Purple
Morphine	Pink/Purple	Purple
Codeine	Violet	Olive green
Acetylcodeine	Purple	Blue
Papaverine	Pink	Purple
Noscapine	No reaction	No reaction
Caffeine	No reaction	No reaction

3.2.4.1 Cobalt isothiocyanate

The reagent is prepared as a 2% w/v solution in water, and is added directly to the test substrate, a positive and negative control.   A blue green colour is obtained from cocaine and all of its surrogates [Table 3.3.].   A number of other compounds also react with this reagent giving so called "false positive" results.

3.2.4.2 Mandelin reagent

The mandelin reagent is applied as for the opiate drugs (section 3.2.3.2). Orange colour reactions are obtained for cocaine and its surrogates.

3.2.4.3 Sodium hydroxide reagent

Also known as the methyl benzoate test, this test examines the presence of a benzoate ester on the drug.   To the test substrate a few drops of 5% sodium hydroxide in methanol are added.   The smell of methyl benzoate ("oil of wintergreen") indicates the presence of a benzoate ester [Table 3.3.].

Very few controlled substances are known to give this combination of results. If such a combination is obtained, it is highly likely that cocaine is present in the seizure.

Table 3.3 Summary of the reactions of cocaine and surrogates
with a number of presumptive test reagents

Drug	Reaction with Reagent		
	Cobalt Isothiocynate	Mandelin Reagent	Sodium Hydroxide
Cocaine	Blue-green	Orange	Methyl benzoate
Ecgonine	Blue-green	Orange	No reaction
Ecgonine methyl ester	Blue-green	Orange	No reaction
Benzoyl ecgonine	Blue-green	Orange	Methyl benzoate

3.2.5 The amphetamines

The amphetamines can be tentatively identified in a sample using the Marquis reagent.   The production of an orange or yellow colour indicates that amphetamine, methamphetamine or another analogue may be present.

3.2.6 The barbiturates

Barbiturates containing drugs are tested using the Dille-Koppanyi reagent. Two solutions are prepared; the first 0.1g cobalt acetate in 10ml 0.2% glacial acetic acid in methanol, the second as 5% iso-propylamine in methanol.   The test is performed by the addition of the first reagent to the test substrate followed by the second.   A blue colour develops if barbiturate drugs are present.

3.2.7 The benzodiazepines

A number of reagents can also be used to test for the presence of benzodiazepine drugs.   These include the Zimmerman Test and the Vitali-Morin test.

3.2.7.1   The Zimmerman test

Two reagents are required, the first 1g 2,4 dinitrobenzene in 100ml methanol, the second 15g potassium hydroxide in 100ml water. A few drops of the first reagent are added to the test substrate, followed by a few drops of the second. A rapidly developing red/purple or pink colour indicates that benzodiazepine derivatives may be present.

It is especially important to perform a positive and negative control with this test because the negative control also gives a colour reaction, but more slowly than if the drug is present.

3.2.7.2   The Vitali-Morin test

Three reagents are required for this test, concentrated nitric acid, acetone and 0.1M potassium hydroxide in methanol. About 0.5ml nitric acid is added to the test substrate, and the mixture dried over a water bath. 5ml of acetone are added, followed by 1ml of the potassium hydroxide solution. A yellow colour develops in the presence of benzodiazepines. Again, it is important to perform both the positive <u>and</u> negative controls since the reagents also produce a colour change, but over a longer time period than if the drug is present.

3.3 Student exercises

When performing these tests, safety gloves, a lab coat and spectacles should be worn. This is because the reagents employed contain strong acids, and strong alkaline solutions. Care should be taken in their use.

3.3.1 <u>Cannabis</u> and its products.

You are provided with a number of samples of <u>Cannabis</u>, including leaf material, resin and oil. Use the Duquenois-Levine and Corinth tests to determine what colour reactions are obtained with these materials.

Repeat the experiment with mace, nutmeg, sage and rosemary.

Q.1 Do you obtain positive results from the non <u>Cannabis</u> material?

Q.2 What does this tell you about the value of the presumptive tests?

3.3.2 The opiate drugs

You are supplied with samples of diamorphine, 6-O-monoacetylmorphine, morphine, acetylcodeine, codeine, papaverine and noscapine. You are also supplied with a sample of raw opium, and a street sample of heroin powder.

> **Q.3** What colour reactions do you get with the various compounds using the Marquis and Mandelin reagents?
>
> **Q.4** Are you able to distinguish between these compounds using these tests?

3.3.3 Cocaine and metabolites

You are supplied with a sample of cocaine, and the related compounds ecgonine, ecgonine methylester and benzoylecgonine. Additionally, you are supplied with a sample of cocaine hydrochloride diluted with a sugar.

> **Q.5** What reactions do you obtain with Mandelin's reagent, cobalt thiocyanate and sodium hydroxide?
>
> **Q.6** How do you explain the smell obtained with sodium hydroxide?

3.3.4 Barbiturate drugs

Using the barbiturate drugs with which you are provided, determine the colour reactions obtained with the Dille Kopanyi reagent. You are also provided with examples of tablets containing barbiturates. Powder the tablet, until you have produced an homogenous mass.

> **Q.7** Can you obtain a positive result from both pure drug and tablet?
>
> **Q.8** Consider whether you could use microscopy and colour tests to positively identify a drug of abuse?
> Explain your answer.

3.4 References

[1] Velapoldi, R.A. and Wicks, S.A.,   Journal of Forensic Sciences 19, 1974, pp.636-656.

[2] Rapid Testing methods of drugs of abuse; manual for use by National Narcotics Laboratories; United Nations, New York, 1988.

4

Thin layer chromatography of drugs of abuse

4.1 Introduction

Having identified the class or classes to which any drugs in the seizure are thought to belong using presumptive testing, the next stage in drug identification and quantification is to examine which members of the class of drug are present in the seizure. This can be achieved through the use of thin layer chromatography (TLC), a technique which is both rapid and cheap, and as such is ideally suited to the screening of samples thought to contain drugs of abuse. The technique uses a separation process that relies upon a layer of material supported on an inert matrix (the so called stationary phase), a solvent moving through this material (the so called mobile phase), and the relative strengths of the interaction between the materials in the sample (the analytes), the mobile phase and the stationary phase.

There are many types of stationary phase available to the drug chemist, but the most common one employed for the analysis of drugs of abuse is silica gel. Silica gel may be modified chemically, in a number of ways, to produce material known as "bonded phases". These materials are expensive, and the analysis times are often long, both of which preclude the use of this material from use in laboratories with a high throughput of case samples. For these reasons such phases will not be considered further.

Silica gel is a polar material, consisting of a number of silanol groups (Si-OH groups) which interact with both the mobile phase and the analytes. The separation of the analytes depends upon the amount of energy dissipated through the interactions of the stationary phase, mobile phase and analytes. If the amount of energy that is dissipated from an interaction between the stationary phase and an analyte is large, relative to that from an interaction between analyte and the mobile phase, then the analyte will be strongly retained, resulting in less movement through the chromatographic bed. The converse is also true; if the interaction between the stationary phase and the analytes is relatively weak, then a larger movement of the analyte through the chromatographic bed will be observed.

The movement of a molecule through the chromatographic bed is measured as the R_f value of the analyte. This is defined as the ratio of the movement of the drug from the origin, relative to the distance of the solvent front from the origin (Fig. 4.1). This means that the R_f value will always lie between 0 and 1 and should only be quoted to two decimal places. Sometimes the R_f value is multiplied by 100 to remove the decimal point.

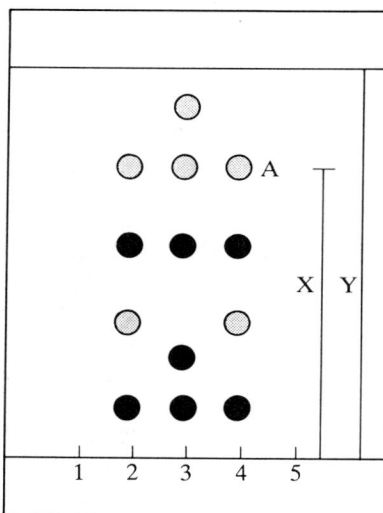

Fig. 4.1 Typical chromatogram after development. Tracks
1,5 = solvent control; 2,4 = positive controls;
3 = sample

$$R_f (A) = \frac{\text{Distance x}}{\text{Distance y}}$$

During application of the sample to the TLC plate, the drug or drug mixture should be fully soluble in the solvent in which it is applied, since particles will cause streaking of the chromatogram. The applied spot should be as concentrated as possible, and hence the more volatile solvents are better for sample application. For an unknown mixture, the amount applied should be that which gives the maximum number of components on visualization but which does not overload the plate. This is usually determined empirically and comes with experience. When applying samples to the TLC plate, tracks should also be included for both negative solvent and positive drug controls. These should be applied in all cases to every chromatographic plate.

Moreover, it is important not to directly compare R_f values on one plate with those on another since these may vary quite considerably.

The solvents employed for TLC on silica gel usually follow an eluotropic series, that is, a series of solvents with eluting power. A typical series is hexane < toluene < dichloroethane < ether < ethanol < methanol < water. Solvents of intermediate power can be made by using appropriate solvent mixtures. The pH of the mobile phase may also be modified by the addition of an acid or base. This will reduce the streaking of the separating components of some mixtures because the modified pH will probably be adjusted to cause charge suppression. The solvents of choice [Table 4.1.] will depend on the class of molecule to be separated, which will have been determined using presumptive testing.

A number of different methods are available for developing a chromatogram. These include single pass, the most common procedure, where the solvent traverses the origin and plate only once; multiple pass where the same solvent is allowed to traverse the plate many times. Multiple pass development can be used to separate stereo-isomers, using a relatively non polar solvent. Over development may lead to loss of resolution, but up to 30 runs may be made.

Stepwise development is used when a mixture is composed of both polar and non polar compounds requiring a polar solvent to develop the plate to half of its height, and a second non polar solvent for full development. Finally, two dimensional development, is where a single sample is applied to the corner of a TLC plate, developed, and then redeveloped in a second solvent, in a direction perpendicular to the first. A complete treatment of the principles and theory of TLC is provided in Braithwaite and Smith [1], although both multiple pass and double development are only used infrequently in drug analysis.

After development, the chromatogram is visualized and the separated components of the mixture can be seen. The TLC plate should first be examined under UV light (254nm and 360nm). Any fluorescence or absorption should be noted. The TLC plate may then be sprayed with one or many of a number of reagents. There are a wide selection of reagents available for drug detection [2,3], but they are often not specific to any one compound. In addition, the sensitivity may vary for different members of the same class of compounds.

Once a compound has been visualised, its Rf value may be calculated, where

$$R_f = \frac{\text{Distance moved by the eluted compound}}{\text{Distance moved by the solvent front}}$$

Table 4.1 Examples of solvent systems commonly employed for TLC [10]
analysis of drugs of abuse

Drug Class	Solvent System (by v/v)		Reference
Cannabis and products	Cyclohexane Di-isopropylether diethylamine	52 40 8	[4]
LSD	Chloroform Methanol	9 1	[5]
Mescaline/Psilocin/ Psilocybin	Methanol Ammonia	100 1.5	[6]
Opiate Drugs	Ethyl Acetate Methanol Ammonia	85 10 5	[7]
	Chloroform Methanol	9 1	[7]
Cocaine & surrogates	Methanol Ammonia	100 1.5	[8]
Barbiturates	Ethyl Acetate Methanol 25% Ammonia	85 10 5	[9]
Amphetamine	Methanol Acetone Ammonia	25 6 0.4	

The R_f will vary depending upon laboratory temperature, purity of the solvents, and mode of development. R_f values alone should not be used to positively identify compounds, but comparison with authentic markers, both in respect of colour reactions **and** R_f values on the same chromatogram, will give an idea of the compound involved.

In addition to calculation of absolute R_f values, two further measures of mobility of an analyte may be employed to identify analytes. The first in the calculation of a relative R_f value, the second the calculation of a corrected R_f value.

Relative R_f values are used to compare the mobility of a compound and that of a standard on a chromatogram. The R_f value of the unknown is calculated, as is that of a standard, and the ratio of R_f values calculated. Whilst absolute R_f values may vary between chromatograms, the relative mobility of compounds in a given solvent remains constant, and may be compared with tabulated data.

Corrected R_f values (hR_f^c values), are obtained by comparing tabulated R_f values of a standard compound against those obtained, in practice in the laboratory. This inclues the origin and solvent front. From this graph, the R_f value obtained is corrected for the mobility of the standards, either by graphical measurement, or by linear interpolation between data points. This is dealt with more fully in de Zeeuw et al [11].

4.2 Spray reagents commonly used in drug analysis

A number of visualisation techniques are commonly used in the analysis of drugs of abuse. These include the use of long and short wavelength UV light and spray reagents. Different classes of drug require different reagents, as tabulated below (Table 4.2). These are only examples, to provide practice in the use of the technique, but do cover the major classes of drug likely to be encountered. TLC spraying should ALWAYS be performed in a flow hood, and suitable protective clothing worn, because of the dangerous nature of TLC spray reagents.

The reagents are prepared as follows:

4.2.1 Fast Blue B

This reagent is freshly prepared as a 1% solution in water. Whilst this reagent gives red, yellow, purple and orange colour reactions with a variety of cannabinoids, it also reacts with a wide variety of phenolic plant products and does not conclusively prove the presence of cannabinoids. However, in combination, the R_f values and colour reactions obtained by it can be used to tentatively identify Cannabis and its products.

4.2.2 Ehrlich's reagent

Ehrlich's reagent for TLC visualisation is prepared by dissolving 1g p-dimethylaminobenzaldehyde in 10ml methanol and adding 10ml concentrated orthophosphoric acid. The reagent is then sprayed directly onto the TLC plate. This reagent can be used to visualise LSD, psilocybin and psilocin, but is also known to react with a wide variety of indole alkaloids.

4.2.3 Acidified potassium iodoplatinate reagent

This reagent provides greater sensitivity than potassium iodoplatinate reagent to which no acid has been added. To prepare the reagent, 0.25g platinic chloride and 5g potassium iodide are dissolved in 100ml water. This is the potassium iodoplatinate reagent. The acidified version is prepared by addition of 2ml of concentrated hydrochloric acid. The reagent is sprayed directly onto the TLC plate. Most controlled substances for which the reagent is used give a deep blue or purple colour reaction.

4.2.4 Dragendorff reagent

Two solutions are mixed to prepare the reagent, the first by mixing 2g bismuth subnitrate, 25ml glacial acetic acid and 100ml of water, the second by dissolving 40g potassium iodide in 100ml water. 10ml of the two solutions, 20ml glacial acetic acid and 100ml of water are mixed to produce the spray reagent, which is sprayed directly onto the TLC plate.

As with the other reagents, this is non specific, yielding an orange colour with a wide variety of alkaloid classes.

4.2.5 Fast Black K reagent

A recently published technique [10] is available for amphetamines. The TLC plate is sprayed with 0.5M sodium hydroxide, and dried with a hair dryer. The chromatogram is then sprayed with 0.5% Fast Black K in water. Amphetamines react almost instantly, giving a red, pink or brown colour reaction.

4.3 Student exercises

A series of exercises follow which illustrate the principles and practice of TLC of drugs of abuse, and some of the associated problems of the technique. The list of methods is not exhaustive, but indicates the types of techniques that are available.

Table 4.2 Examples of commonly used methods for visualisation of drugs separated by TLC
*The mercuric chloride-diphenylcarbazone reagent should not be employed because of the extremely toxic nature of mercuric compounds.

Drug Class	Visualisation technique employed	Colour obtained with spray reagent
<u>Cannabis</u> and products	UV light 254nm Fast Blue B	absorbs UV light Orange/Yellow/Red
LSD	UV light 254nm 360nm Ehrlich's reagent	absorbs UV light Fluorescent Blue/Purple
Opiate drugs	UV light 254nm Acidified potassium iodoplatinate Dragendorff reagent	absorbs UV light Blue/Purple Orange

Cocaine	UV light 254nm
	
	Acidified potassium
	  iodoplatinate
		some absorb UV
		  light
		Blue/Purple

Barbiturates*	UV light 254nm	absorb UV light

Benzodiazepines	UV light 254nm
	1M H_2SO_4/heat/
	  UV light 366nm
	Acidified potassium
	  iodoplatinate
		absorbs UV light
		fluoresce
		
		Blue/Purple

Amphetamines	UV light 254nm 0.5M Sodium hydroxide 0.5% Fast Black K	 Red/Brown/Pink

4.3.1 <u>Cannabis</u> and its products

You are provided with samples of Δ^8-THC, Δ^9-THC, cannabidiol and cannabinol in addition to <u>Cannabis</u> plant material and some <u>Cannabis</u> resin. The plant material may conveniently be ground to a find powder in a pestle and mortar, and extracted with ethanol, or diethyl ether, again either using cold extraction or a soxhlet.   Examine the materials by thin layer chromatography, and calculate the R_f values of the compounds of interest.
     Use the TLC system described below.

Stationary phase:	Silica gel G, 0.2 mm thickness
Solvent System:	Cyclohexane/di-isopropyl ether/diethylamine (52/40/8 v /v /v)
Detection:	U.V. Light 254 nm and 360 nm. Fast Blue B (Section 4.2.1)

Q.1 What are the R_f values of the cannabinoids that can be
     detected using this method?   How else may the compounds
     be differentiated?

Q.2 What difficulties do you encounter in compound identification?

Q.3 As your solution, prepared from the <u>Cannabis</u> products,
     becomes older, do you notice any change in composition?
     What does this tell you?

4.3.2 <u>The opiate drugs</u>

You are provided with examples of opiate drugs, some adulterants and examples of street seizures. Examine these compounds by TLC, using the following systems.   Solutions at 1mg/ml in methanol are suitable for application to the TLC plate.

Stationary Phase:	Silica Gel G, 0.2 mm thickness
Solvent System:	Ethyl acetate/Methanol/Ammonia (85/10/5 v/v/v) Chloroform/methanol (9/1 v/v)

Detection: UV light at 254 and 360 nm
                              Acidified potassium iodoplatinate (Section 4.2.3)

Q.4 Why do you think that the ammonia is included in the
solvent system?

Q.5 How do you account for the R_f values that you observe?

Q.6 How do you account for any differences that you observe
between your data and tabulated data [Table 10.2.]?

Q.7 What does this tell you about examining unknown compounds
by TLC alone?

Q.8 What happens as your solution of diamorphine ages?

4.3.3 Cocaine and surrogates

You are provided with samples of cocaine, ecgonine, ecgonine methyl ester
and benzoylecgonine. Using the following TLC system, determine the R_f
values of these compounds.

Stationary Phase: Silica gel G, 0.2 mm thickness

Solvent System: Methanol / Ammonia
                         (100 / 1.5 v/v)

Detection System: Acidified potassium iodoplatinate (Section 4.2.3)

Q.9 How do you explain the R_f values observed for cocaine and
related compounds, in terms of their structures, and the
separation process?

4.3.4 Barbiturate drugs

You are provided with examples of barbiturate drugs and some barbiturate
containing tablets. Powder the tablets and shake the powder with an

appropriate volume of methanol, filter or centrifuge the mixture to obtain a clear solution and reduce the volume to about $100\mu l$ in a stream of nitrogen gas.   Using this extract and the methanolic solutions of standard barbiturates make an assessment of the barbiturate content of the tablets.

Stationary Phase: Silica Gel GF_{254}, thickness
0.2 mm.

Solvent System: Ethyl Acetate/Methanol/25% Ammonia
85 / 10 / 5 v/v/v
Chloroform/Acetone 80 / 20 v / v

Detection: U.V. Light at 254 nm and 360 nm

NB: the detection method employing mercuric compounds, recommended by the United Nations **SHOULD NOT BE USED** because of the hazardous nature of the mercuric compounds.

Q.10 What is the principle problem with this technique?

Q.11 Are you able to determine which compounds are present in the tablets?

Q.12 How do you explain the R_f values that you observe?

4.4 References

[1]   Braithwaite, A. and Smith, F. J., Chromatographic Methods, 4th edition, Chapman and Hall, 1982.

[2]   Smith, I.,   Chromatographic and Electrophoretic Techniques, vol. I., Heinemann, Berlin, 1969

[3]   Wagner, H., Bladt, S. and Zgainski, E.M.,   Plant Drug Analysis, Springer Verlag, Berlin, 1983.

[4]   Recommended methods for testing Cannabis, United Nations Division of Narcotic Drugs, New York, 1988.

[5]   Recommended methods for testing Lysergide (LSD),   United Nations Division of Narcotic Drugs, New York, 1988.

[6] Recommended method for testing peyote cactus (mescal buttons)/Mescaline and psilocybe mushrooms/psilocybin,   United Nations Division of Narcotic Drugs, New York, 1988.

[7] Recommended methods for testing Heroin,   United Nations Division of Narcotic Drugs, New York, 1988.

[8] Recommended methods for testing cocaine,   United Nations Division of Narcotic Drugs, New York, 1988.

[9] Recommended methods for testing barbiturate derivatives under international control,   United Nations Division of Narcotic Drugs, New York, 1988.

[10] Munro, C. H. and White, P. C.   "Evaluation of diazonium salts as visualisation reagents for thin layer chromatographic characterisation of amphetamines",   Journal of the Forensic Science Society,   In press.

[11] de Zeeuw, R. Aa, Frauke, J.P., Degal, F., Machbert, G., Schutz, H. and Wijsbeek, J.,   Thin layer chromatographic R_f values of toxicologically relevant substances on standardised systems.   VCH mbH, Weinheim, 1992.

5

Gas chromatography of drugs of abuse

5.1 Introduction

Having tentatively identified the members of the drug classes thought to be present in the sample by presumptive testing and thin layer chromatography, the next stage in their analysis is the confirmation of their identity, and their quantification. One technique which can be employed to do this is gas chromatography. This technique relies on drugs or their derivatives being volatile, thermally stable and the ability of the system to separate them in the vapour phase.

The gas chromatograph involves a system for loading the sample onto the column (the injection port), a column, in an oven at a controlled temperature, on which the separations occur and a detection system (Fig. 5.1.).

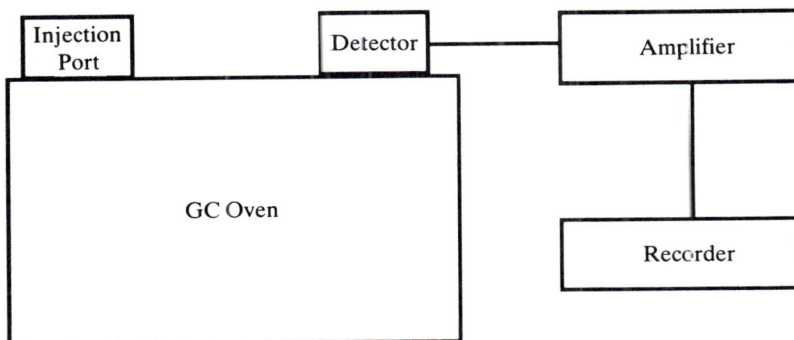

Fig. 5.1 Components of a gas chromatograph

The sample is inserted into the gas chromatograph through the injection port, where it is volatilised, and swept onto the column by the flow of the carrier gas.   If the sample is not very concentrated, as for toxicological samples, it may be necessary to chromatograph the whole sample.   More often, however, if the whole sample is introduced onto a capillary column (the type of column most commonly employed), the column will become saturated with sample.   For this reason, the sample must be "split", so that only a controlled amount is swept onto the column.   This is achieved through a "split valve", a part of the injection port especially designed for this purpose (Fig. 5.2).   The split ratio is most commonly measured as the ratio of gas exiting the split valve, divided by the carrier gas flow rate.   If, for example, the flow rate through the split valve was 50ml/min and the column flow rate 1ml/min, the split ratio would be 50:1.

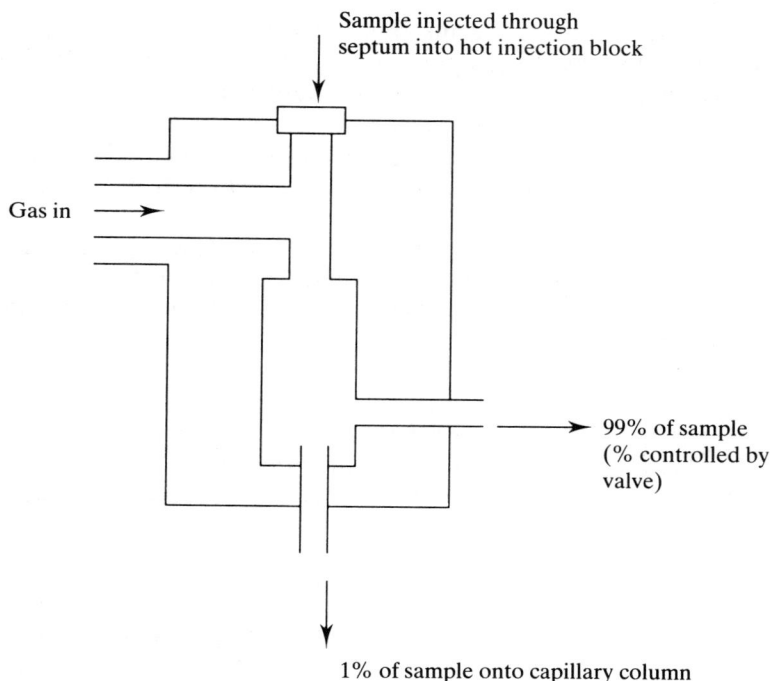

Sample injected through
septum into hot injection block

Gas in ⟶

99% of sample
(% controlled by
valve)

1% of sample onto capillary column

Fig. 5.2 Schematic representation of construction of a split valve

Once the components of the sample are swept onto the column, the separation process begins.   The separation of the components of a mixture is achieved through a partitioning process between the carrier gas (the mobile phase) and the liquid phase (the stationary phase). Since in gas chromatographic drug analysis the stationary phase is a liquid it must be coated onto a support. This support is either the wall of the column itself, as in capillary gas chromatography, or a support material, as in packed column gas chromatography (Fig. 5.3.).

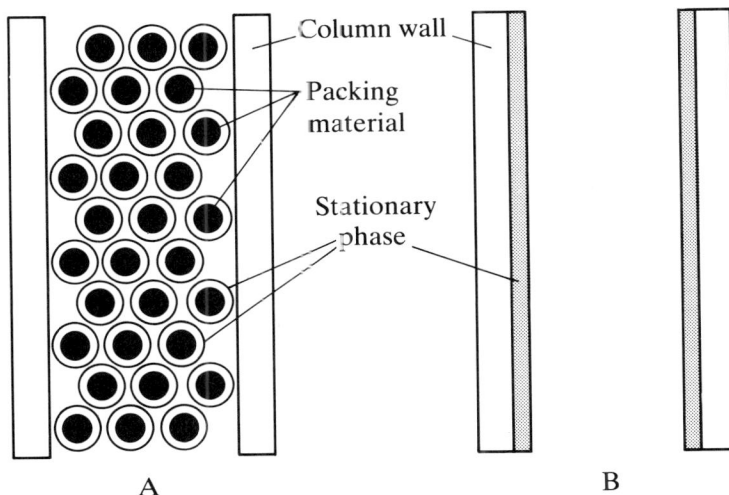

Fig. 5.3 Coating of stationary phase onto support in (A) a packed column, and (B) a capillary column

In general, capillary column gas chromatography has largely superseded packed column chromatography, but packed column chromatography is still employed in some laboratories. Packing materials are generally diatomaceous earths which are prepared by chemical and/or physical treatment of materials from geological sources. The packing density, surface area per unit mass, and mesh size of the particles of which the packing material is composed will all influence the quality of the separation. For general purpose chromatography of commonly occurring drugs, particle sizes of about 80-120 mesh are used.

There are a number of stationary phases available to the drug analyst, which possess a variety of polarities; a number of these are listed in Table 5.1. Polar stationary phases are used to separate polar compounds. and non polar columns to separate non polar compounds. For screening purposes, OV-1, BP-1, OV-17, or their equivalents, are sufficient for most drug analyses. It is the differences in polarities of the drugs, their molecular weights and their volatility, on which the separation process is based.

The resolution of drugs achieved using a packed column is not normally so good as that of a capillary column. The increased resolution of the latter is due to the elimination of eddy diffusion which occurs in a packed system as a result of the multitude of pathways between the particles taken by the mobile phase and the gaseous analytes. Resistance to mass transfer is also reduced in both the gas and the liquid phases in capillary columns because of their small internal diameters and the controlled film thickness.   These considerations are dealt with more fully in chromatography texts [1].

The differences between SE-30 and OV-1/OV-101 lie in the chain length of the polymethylsiloxanes. The longer the chain length, the higher the maximum possible operating temperature. This is because the longer the chain length, the higher the temperature required to cause the stationary phase to "bleed".

Table 5.1 Examples of stationary phases used in drug analysis

Stationary Phase	Max. Temp °C	Phase Type	Uses
Apiezon L	300	Hydrocarbon grease	Barbiturates Amphetamines
SE-30	300	Dimethyl-silicone	General purpose
OV-1	350	Dimethyl-silicone	General purpose
OV-101	350	Dimethyl-silicone	General purpose
OV-17	350	Phenylmethyl-silicone	General purpose
Carbowax 20M	225	Polyethylene glycol	Alcohols

OV-17 is more polar than OV-1 due to the presence of the phenyl moiety of the stationary phase. This allows the interaction of the pi electrons from the stationary phase and any unsaturated system in the solute molecule. Carbowax is more polar still, due to the many hydroxide moieties found in polyethylene glycol.

The separation process is achieved through partition of the analytes between the mobile phase and the stationary phase.   Once the sample is introduced into the GC column, the analytes partition into the stationary phase, Fig. 5.4 (a, b).   The strength of the interaction between the stationary phase and analytes depends upon the structure of both, but in general, on OV-1 and OV-17, the more lipophilic compounds partition preferentially into the stationary phase, whilst the more polar compounds, at a given temperature, re-enter the gas phase more easily (Fig. 5.4c).   Following this, the more lipophilic molecules also re-enter the gas phase (Fig. 5.4d).   The relative mobility of molecules is also determined by their molecular weights - the higher the molecular weight, the greater the residence time in the stationary phase.

Two major types of temperature regimes are employed in GC-isothermal chromatography, using the same temperature throughout the analysis, and temperature programmed chromatography, where the temperature of the column increases as the analysis progresses.   This latter type of temperature control is particularly useful when analytes with a wide range of polarities and volatilities are to be analysed.

Whilst some drugs will chromatograph satisfactorily without chemical modification, for example, diamorphine and cocaine, their surrogates and breakdown products may not chromatograph well or be well resolved due to their polarity (Fig. 5.5.).   Other compounds, such as the cannabinolic acids may be thermally unstable.   For these reasons, chemical derivatisation is often used in drug analysis.   This process increases thermal stability of the drug, and improves chromatographic behaviour by derivatising polar functional groups, for example, hydroxide and amino moieties (Fig. 5.6.).

Two principal types of derivatisation are employed.   The first involves pre-column derivatisation, as for the trimethylsilylation of opiate, cocaine and cannabinoid based drugs.   A small amount of the drug is placed in a derivatisation vial, and the solvent evaporated under nitrogen.   The residue is then derivatised directly, with N,0-bistrimethylsilylacetamide (BSA).   The reaction occurs almost instantly (Fig. 5.7.).

However, the sample must not contain water since this will hydrolyse the trimethylsilylated derivative.   When the analysis is performed, a control of solvent residue derivatised with BSA should be chromatographed, to ensure that any drugs present are not confused by products of the derivatisation process.

The second form of derivatisation is known as "on-column" or "flash alkylation".   This is used on, for example, barbiturates, to N-methylate the nitrogen atoms (Fig. 5.8).   A methanolic solution of the drug is place in a derivatisation vial and dried under nitrogen.   A few drops of the reagent, trimethylanilium hydroxide, as a 0.2M solution in methanol are added and mixed with the drug residue and the sample injected into the gas chromatograph.   In this case the derivatisation process occurs in the injection block.

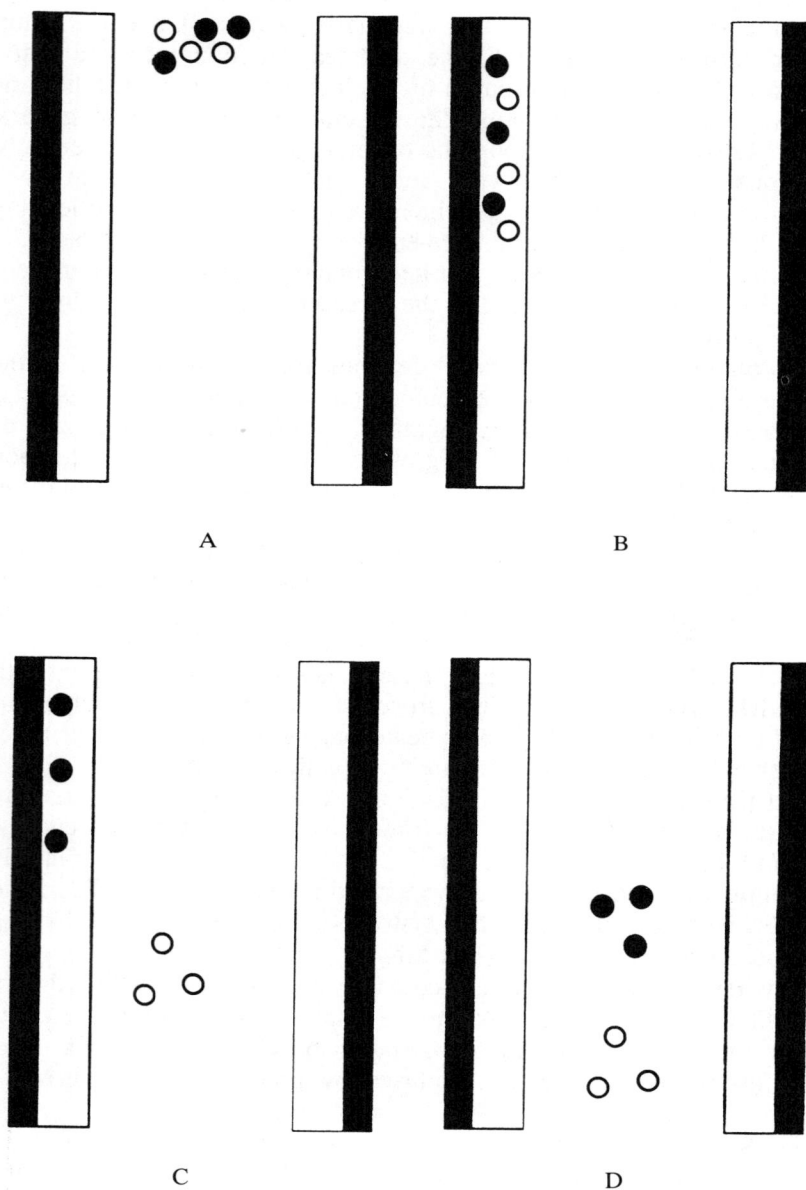

Fig. 5.4 Representation of a separation step for 2 analytes, using
capillary gas chromatography

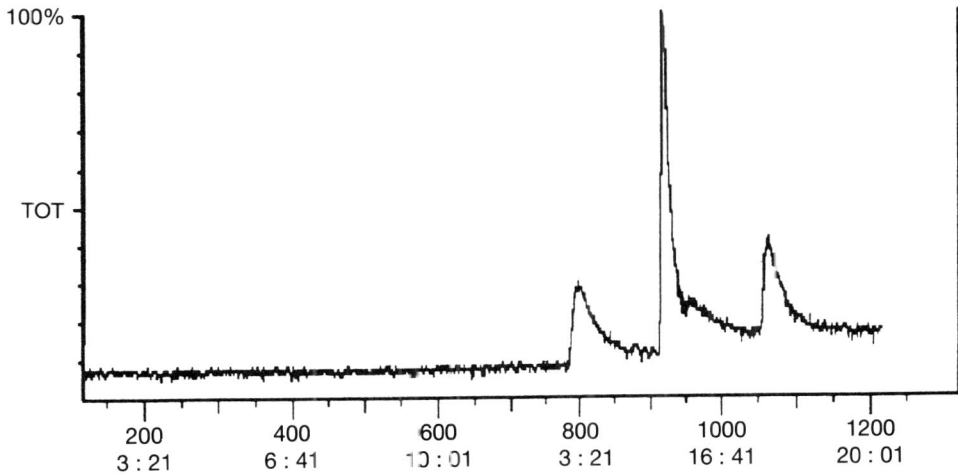

Fig. 5.5 Gas chromatogram showing poor chromatography of opiate drugs on a
BP-5 capillary column prior to derivatisation with N,0-bistrimethylsilyl
acetamide

Whenever derivatisation is employed, care should be taken that protective clothing is worn and that the process is carried out in a flow hood which draws air <u>away</u> from the operator, since the reagents will often react with certain components of body tissues.

Once the components of the mixture have been introduced onto the column and are separated, they must then be detected as they leave the column in sequence. A number of detector systems are available which are of use for the analysis of drugs. These include the flame ionization detector, the nitrogen phosphorus detector, the electron capture detector, and the mass spectrometer. The last of these is discussed in greater detail in chapter 9.

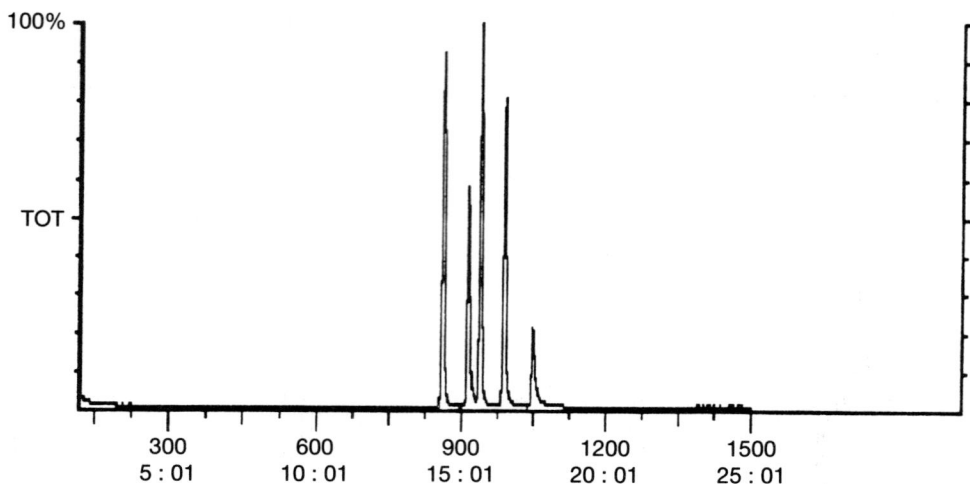

Fig. 5.6 Gas chromatographic separation of opiate drugs on a BP-5
capillary column after derivatisation with N,0-bistrimethylsilylacetamide
(elution order: codeine, acetylcodeine, morphine, 6-0-monacetylmorphine,
diamorphine)

Flame ionisation detectors (Fig. 5.9) are the most commonly used detector, possessing a wide range of linear response to the presence of organic molecules over seven orders of magnitude of concentration. The detector is essentially a hydrogen-air flame which burns below a detector electrode maintained at a voltage of about 150V with respect to the flame jet. The hydrogen is introduced into the column eluent, and enters the jet with the eluent, where the mixture burns.   As organic compounds enter the flame, which contains H˙, O˙ and HO˙ radicals, they undergo a complex series of reactions which for those components containing carbon and hydrogen result in CH˙ radicals. These radicals interact with oxygen radicals to produce positively charged CHO^+ ions and an electron. The former are attracted to the charged electrode where they are discharged. By means of an electrometer which converts charge to current and an amplifier, an electronic signal is produced which is usually recorded on a chart recorder or an integrator, as a peak. In this way, the signal produced is related to the carbon number of the molecule, although other functional groups, for example alcohol, amino and carbonyl groups, will have an effect. A further advantage of this type of detector is that it does not respond to some solvents, for example carbon disulphide and water.

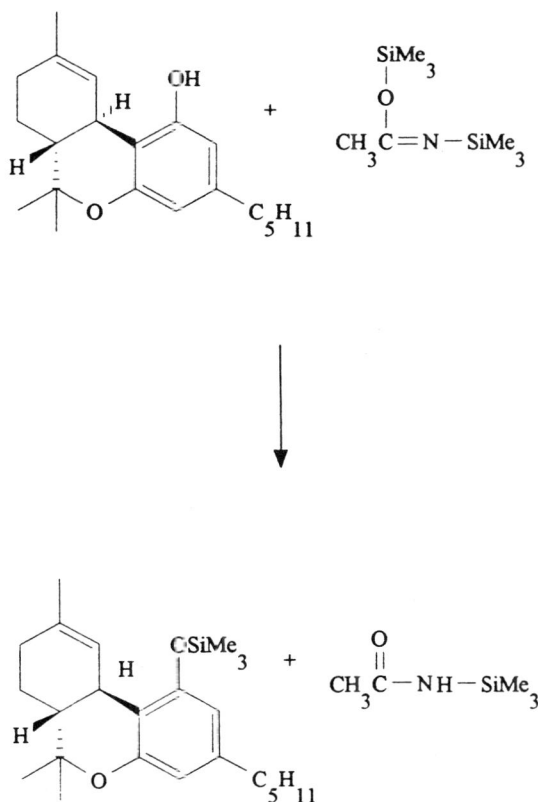

Fig. 5.7 Derivatisation of Δ^9-tetrahydrocannabinol with
N,0-bistrimethylsilylacetamide

FID detectors are the most frequently employed detectors for routine screening and drug analysis.   However, other more selective detectors are available, where greater selectivity is required.   Most drugs of abuse contain nitrogen, and the selectivity towards these of the flame ionization detector can be improved by making the detector nitrogen and phosphorus specific using the so called nitrogen phosphorus (NPD) detector.

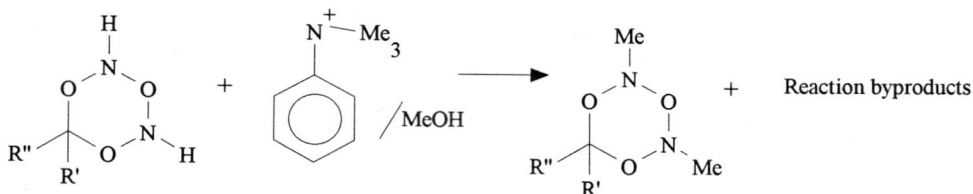

Fig. 5.8 Derivatisation of a barbiturate with 0.2M trimethylanilium hydroxide in methanol

A bead of an alkali metal salt is placed between the FID tip and the collector electrode (Fig. 5.10.). A number of alkali salts can be used. For example, caesium bromide is used for organophosphorus compounds, rubidium chloride or rubidium sulphate for organonitrogen compounds and potassium chloride or potassium carbonate for halogenated compounds.

In the NPD mode, the hydrogen burns at the surface of the bead producing a plasma. To stabilize this plasma requires very careful control over the flame and carrier gases, particularly the former, and especially the hydrogen flow. This is because the detector depends upon the concentration of the alkali ion. The need for such careful control presents the major disadvantage when compared to FID, but this is sometimes outweighed by the increase in detector selectivity.

When burnt in either the flame situated below the bead, or in the plasma itself, nitrogen containing compounds produce CN· radicals, which then react with the alkali metal ions to produce cyanide ions. These ions migrate to the collector electrode, where they are discharged, producing the signal. The NPD detector is about 50 times more sensitive to nitrogen containing compounds relative to the FID and gives a linear response over six orders of magnitude. Whilst the added sensitivity to nitrogen containing compounds is a bonus this detector type has vastly reduced sensitivity to compounds containing only carbon, hydrogen and oxygen. This means that there are usually few interfering peaks from source, and this is seen particularly in a reduction of the "solvent front" response.

Electron capture detectors use the changes in electrical conductivity of gases in an ionisation chamber caused by the presence of electron acceptor compounds bearing electron negative groups in the GC eluate. The detector cell consists of a β particle emitter, usually ^{63}Ni, and a collector electrode. The high energy electrons (β particles) interact with the nitrogen carrier gas and in

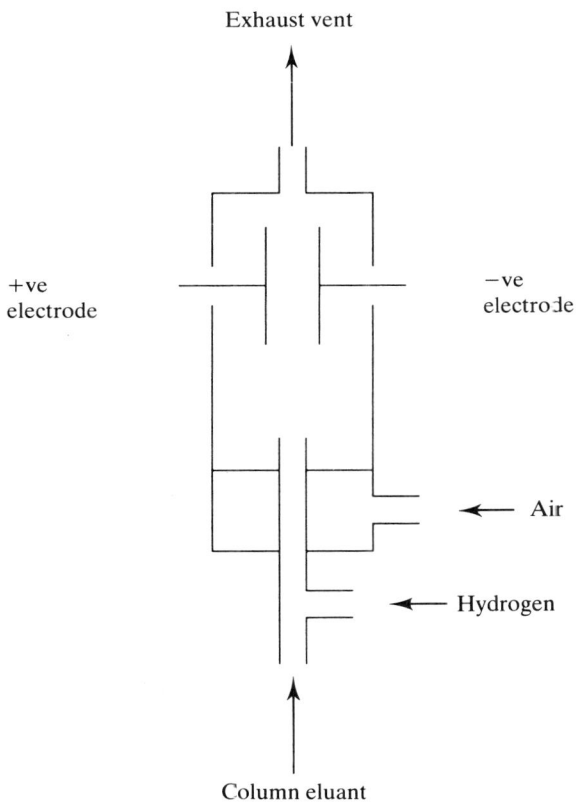

Fig. 5.9 Schematic diagram of a flame ionisation detector

the process, produce a baseline current. Molecules with high electron affinities catch these high energy electrons as the compounds pass through the detector (Fig. 5.11.).

This interrupts the baseline current, and the resulting change in voltage is used to produce a chromatogram. The disadvantage of this method is that it does not have a wide linear range - being only at best 4 orders of magnitude

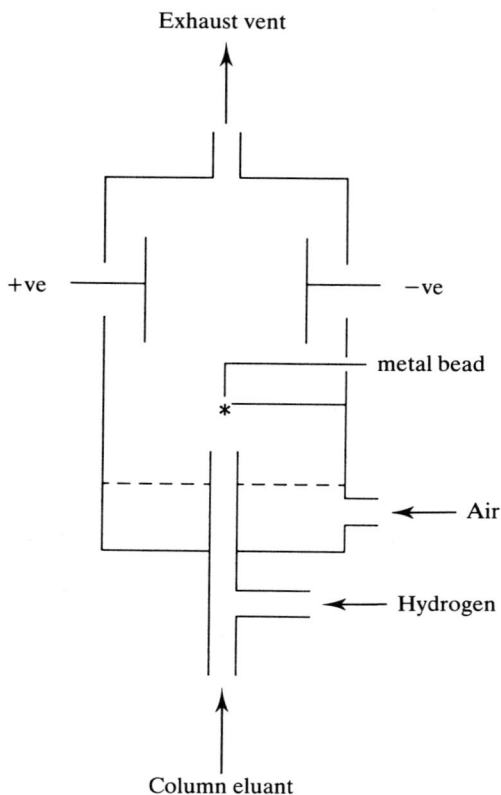

Fig. 5.10 Schematic diagram of a nitrogen phosphorus detector

of concentration. The sequence of reactions which take place within the detector can be summarised as follows:

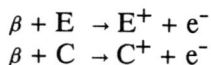

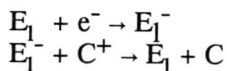

$$\beta + E \rightarrow E^+ + e^-$$
$$\beta + C \rightarrow C^+ + e^-$$

$$E_1 + e^- \rightarrow E_1^-$$
$$E_1^- + C^+ \rightarrow E_1 + C$$

E = eluate
C = carrier gas
E_1 = electron capturing compound
Net effect is loss of an electron i.e. current reduction

It is important to realise that the electron capture detector is a much more specific detector than either the FID or the NPD detector. In practice it is only useful for those drugs containing halogen, nitro and/or carbonyl moieties. However, it should be borne in mind that many drugs can be derivatised with halogen bearing compounds prior to analysis, for example, the pentafluorobutyryl derivative of morphine.

A mass spectrometer can also be attached to a gas chromatograph to act as a sophisticated detector. Many systems operate using the electron impact mode which can be used to display a total ion current chromatogram, similar to a standard GC trace from an FID, NPD or ECD detector. In addition, molecular ions and fragmentation pathways can also be observed, which are important for comparison of samples with library data with a view to identifying unknown peaks. Gas chromatography-mass spectrometry is dealt with more fully in Chapter 9.

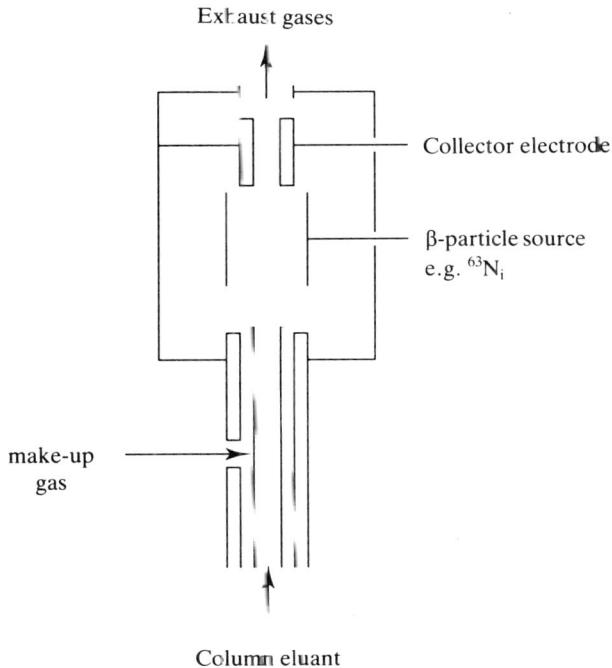

Exhaust gases

Collector electrode

β-particle source
e.g. $^{63}N_i$

make-up
gas

Column eluant

Fig. 5.11 Schematic diagram of an electron capture detector

5.2 Interpretation of chromatographic data

A chromatogram gives a series of peaks indicating that different compounds have been eluted from the system. These peaks can be used to both identify and quantify any drugs present.

5.2.1 Standard mixes

Before a GC system is employed for the analysis of drugs of abuse, it should first be checked for performance. Columns are often supplied with standard mixes of compounds with different chromatographic properties. A typical mixture employed for a column used in drug analysis, such as a PV-1, or its equivalent, BP-1, column, will contain one or more n-alkanes, a phenolic compound, an alkylamine and a long chain n-alcohol. If the column is performing properly, all should give good peak shapes.

Additionally, at the start and end of each analytical sequence, a mixture of positive controls should be chromatographed, to ensure that the column is performing properly.   Between each sample analysed an injection of pure solvent should be made to ensure that there is no between injection carry over.

5.2.2 Retention time

Sample peaks can be tentatively assigned to compounds by comparison of their retention times with those of standard drugs analysed under identical conditions. However, it is quite possible that mis-identification may be achieved due to the problems of co-elution; that is, two or more compounds having the same retention time.

5.2.3 Kovat's indices

If a standard is not available, then tentative identification of a drug in a sample can be made through the use of Kovat's Indices [2], and an isothermal chromatographic system. A series of n-alkanes (for FID) or alkylamines (for NPD) are chromatographed. An alkane or alkylamine with retention time longer, and another with retention time shorter than that of the unknown compound are chromatographed, and the retention times inserted into the Kovat's Index Equation:

$$KI = 100 \left[n \left(\frac{\log t_{r(u)} - \log t_{r(x)}}{\log t_{r(x+n)} - \log t_{r(x)}} \right) + x \right]$$

where

$t_{r(u)}$ = retention time of the unknown,

$t_{r(x)}$ = retention time of the smaller alkane,

$t_{r(x+n)}$ = retention time of the larger alkane,

n = difference in the carbon number of the alkanes,

x = carbon number of the smaller alkane.

This is illustrated below for diamorphine, using a 25m capillary column, 0.5μm layer thickness.

t_r (diamorphine) = 1293 seconds

t_r ($C_{22}H_{46}$) = 361 seconds

t_r ($C_{28}H_{58}$) = 1486 seconds

n = 6

x = 22.

Substituting these values into the equation gives :

$$KI = 100 \left[6 \left(\frac{3.112 - 2.558}{3.172 - 2.558} \right) + 22 \right]$$

= 2741.

In practice, the alkanes do not have to be consecutive in the series. In this example, C_{22} and C_{28} have been used, but so could C_{26} and C_{28}. Once the index is calculated, the value obtained is compared to tabulated values, obtained on the same GC system. It is important to note here that the Kovat's Index of a drug will differ in value from that of any derivative of the drug. Kovat's Index values will also vary between column types. In reality, whilst this may be used for bulk samples, and for trace samples where no

GC-MS is available, in most laboratories, GC-MS techniques would be applied directly, to trace samples.

The Kovats equation given above only applies for isothermal operation for temperature programming the direct retention times and not the logarithm of the retention time should be employed. Moreover, when comparing values of Kovat's Indices with those of a data base, to make sure that the unknown compound is amongst those in the data base a window of 50 Kovat's Indices should be employed.

Whilst Kovat's Indices and/or retention times provide clues as to the identify of a drug, this data alone cannot be relied upon as absolute proof of identity. This can best be provided using gas chromatography-mass, spectrometry, discussed in Chapter 9.

5.2.4 Quantification

In addition to the identification of components of a mixture, gas chromatography can also be used to estimate the amount of each component. If derivatisation is to be used, then all precautions should be taken to ensure that it is complete.

For HPLC, a fixed volume may be accurately injected into the analytical system, but the same degree of accuracy of injected volume cannot be achieved with a gas chromatograph, except where certain types of autosamplers are employed. This problem is overcome by means of an internal standard which is a compound chosen so that it chromatographs clear of the compounds of interest, but is usually of the same generic type. This is illustrated below for a sample thought to contain diamorphine.

A calibration curve is constructed by injecting approximately the same volume of a number of standard solutions which contain exactly the same concentrations of internal standard but different concentrations of the drug to be analysed. A concentration higher than that expected, and another that is lower than would be expected in street samples should always be included to ensure that the sample concentration will be on the linear range of the graph. The data is conveniently recorded in tabulated form. The ratio of peak heights of the sample to internal standard is calculated and a regression equation calculated [Table 5.2.]. A graph of relative peak height of diamorphine to the internal standard against the concentration of sample injected is constructed, as illustrated (Fig. 5.12.). The suspected sample is then chromatographed in the presence of the same concentration of internal standard.

From the graph of relative peak area for the standards and the relative peak areas of the sample, the concentration of the sample can either be read graphically, or calculated from a rearranged form of the regression equation, calculated using the calibration data.

Table 5.2 Data obtained for calibrating a GC for Diamorphine, using $C_{28}H_{58}$ as an internal standard

Peak height diamorphine	Peak height C-28	Ratio of peak heights (y)	Diamorphine concentration (x)	(x²)	Ratio x concentration (xy)
122.00	1,102.00	0.11	0.25	0.063	0.03
130.00	1,110.00	0.12	0.25	0.063	0.03
127.00	1,107.00	0.11	0.25	0.063	0.03
350.00	1,512.00	0.23	0.50	0.250	0.12
376.00	1,480.00	0.25	0.50	0.250	0.13
370.00	1,500.00	0.25	0.50	0.250	0.12
750.00	1,542.00	0.49	1.00	1.000	0.49
770.00	1,530.00	0.50	1.00	1.000	0.50
850.00	1,540.00	0.55	1.00	1.000	0.55
1,423.00	1,634.00	0.87	1.50	2.250	1.31
1,400.00	1,620.00	0.86	1.50	2.250	1.30
1,275.00	1,550.00	0.82	1.50	2.250	1.23
2,413.00	2,159.00	1.12	2.00	4.000	2.24
2,323.00	2,050.00	1.13	2.00	4.000	2.27
2,187.00	2,020.00	1.08	2.00	4.000	2.17
1,750.00	1,261.00	1.39	2.50	6,250	3.47
3,290.00	2,520.00	1.31	2.50	6.250	3.26
3,450.00	2,540.00	1.36	2.50	6.250	3.40
Totals 23,356.00	29,777.00	12.56	23.25	41.44	22.63

When performing the injection sequence it is important to start the injection at the lowest concentration and work towards the most concentrated standard. Each standard should be chromatographed twice, and all injections should be separated by a blank injection, of solvent alone, to ensure that no sample carryover has occurred. It should also be noted that at least 12 data points are required for the calculation of a reliable regression equation. Whilst this seems excessive, with autosamplers this should not present a problem.

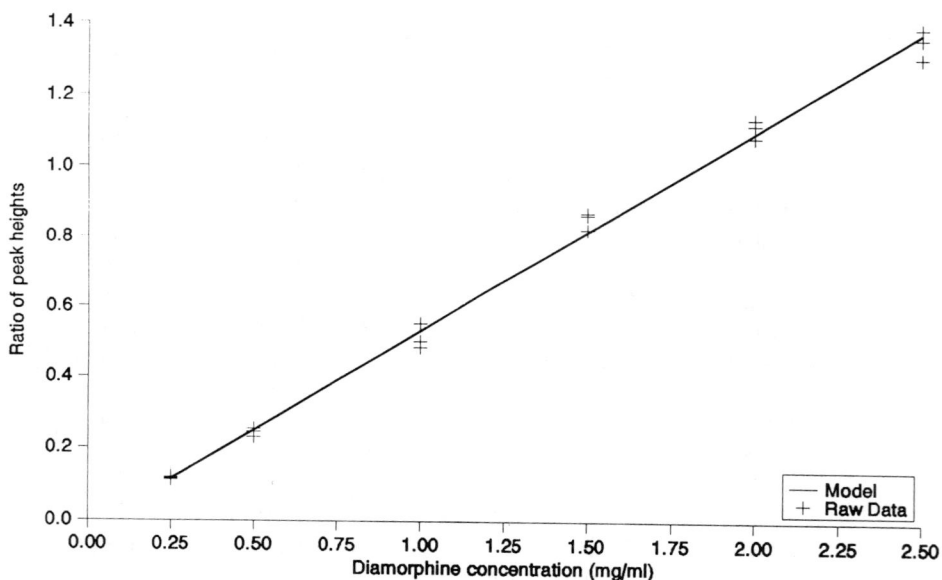

$Y = 0.56 X - 0.03$

Fig. 5.12 Calibration curve for diamorphine, using an FID, and the GC
conditions described (Section 5.3.3) and $C_{28}H_{58}$ as an
internal standard

The regression equation for ratio of peak heights against concentration
can be assumed to be of the form $y = mx + c$, where y is ratio of the peak
heights, m is the gradient of the line, x is the diamorphine concentration and c
is a constant.   The values of m and c can be solved using the equations

$$\Sigma y = m\Sigma x + nc \quad (1)$$
and
$$\Sigma xy = m\Sigma x^2 + c\Sigma x \quad (2)$$

where n is the number of data points (in this example $n = 18$).

The data obtained in this example gives the following values:

$$12.56 \quad = \quad 23.25m + 18C \qquad (3)$$
$$22.63 \quad = \quad 41.44m + 23.25C \qquad (4)$$

Solving equations (3) and (4) for m and c gives values m = 0.56, c = 0.03. This gives the equation y = 0.56x - 0.03, which can be rearranged to give a diamorphine concentration from a peak height ratio, thus

$$\frac{y + 0.03}{0.56} = \text{Diamorphine concentration}$$

5.3 Student exercises

5.3.1 Kovat's indices and opiate drugs

The opiate drugs illustrate well the use of Kovat's Indices. Using the following gas chromatographic system, calculate Kovat's Indices for diamorphine, monoacetylmorphine, morphine, acetylcodeine and codeine. In addition investigate whether the compounds give the same values as a mixture. Useful n-alkanes are C-20 and C-28.

Detector:	Flame Ionization Detector (FID)
Column:	OV-1, 0.5 μm thickness, 25M
Split ratio:	60:1
Carrier gas flow rate:	1 ml/min.
Injection temp:	300°C
Detector temp:	300°C
Column temp:	240°C

Q.1 How do you explain the retention times that you observe for for these compounds?

Q.2 How do the values for the Kovat's Indices for the drugs compare with those in the literature?

Q.3 What problems do you envisage when using this method to investigate illicit samples of diacetylmorphine and "heroin"?

5.3.2 Gas chromatography of opiate drugs using temperature programming

Co-elution of compounds can, in part, be overcome by the use of temperature programming.

Repeat the above experiment using the following temperature programme :

Injection Temp:	300°C
Detector Temp:	300°C
Column Temp:	150 - 280°C @ 8°C/min.

> **Q.4** What kind of separation do you now achieve.   Do you still have problems with some compounds?

5.3.3 Gas chromatography of opiate drugs using derivatisation and temperature programming

The following gas chromatographic method has been employed to separate opiate drugs and their common adulterants.

Detector:	FID
Column:	OV-1, 0.5μm layer thickness, 25M
Split ratio:	50:1
Carrier gas flow rate:	1ml/min
Injection temp:	250°C
Detector temp:	280°C
Temp. Programme:	150°C for 2 minutes
	150°C - 280°C @ 9°C/min
	280°C for 2 minutes
Derivatisation reagent:	N,0-bistrimethylsilylacetamide

Determine the type of separation that you can achieve.

> **Q.5** How do you explain in chemical terms, the improved chromatography of morphine?

5.3.4 Cannabinoids

There are now at least 69 cannabinoids known to be produced by <u>Cannabis</u> <u>sativa</u>.

It should be noted at the outset that gas chromatography of the cannabinoids may present problems in that if derivatisation is not employed; the carboxylic acids will thermally degrade.

Using the method described below, it is possible to separate these compounds, as their BSA derivatives, using capillary gas chromatography.

Detector:	FID
Column:	OV-1, 0.5 μm layer thickness, 25M
Split ratio:	50:1
Carrier Gas flow rate:	1 ml/min
Injection Temp:	300°C
Detector Temp:	300°C
Temp Programme:	170°C for 2 minutes,
	170°C - 280°C at 4°c/min.
	280°C for 2 minutes

Using this system, chromatograph the cannabinoids alone and as a mixture.

You are provided with plant material of <u>Cannabis</u> <u>sativa</u>, and an example of Lebanese Gold cannabis resin. Extract each with diethylether, and after decanting the supernatant, blow to dryness under nitrogen. Derivatise with BSA, and chromatograph under the previously described conditions.

> **Q.6** Are you able to identify any components of the mixture, what problems do you encounter and how may they be overcome?

5.3.5 Cocaine and surrogates

Whilst cocaine itself will chromatograph satisfactorily, its surrogates benzoyl ecgonine, ecgonine and ecgonine methyl ester require derivatisation. Using the method below, obtain chromatographic data for these compounds.

Detector:	FID
Column:	OV-1, 0.5μm layer thickness 25M
Split ratio:	50:1
Carrier gas flow rate:	1ml/min
Injection Temp:	300°C
Detector Temp:	300°C
Temp Programme:	170°C for 2 minutes
	170°C - 280°C @ 16°C/min
	280°C for 2 minutes
Derivatising reagent:	N,0-bistrimethylsilylacetamide

Q.7 Why do the surrogates require derivatisation?

You are provided with a sample simulating a street seizure of cocaine. Attempt to identify the components of the mixture, and quantify the cocaine present.

5.3.6 The barbiturate drugs

There are a wide variety of barbiturate drugs available. Using the on column derivatisation method described below, separate the examples with which you are provided.

Detector:	FID
Column:	OV-1, 0.5μm layer thickness, 25M
Split ratio:	50:1
Carrier gas flow rate:	1ml/min
Injection temp:	275°C
Detector temp:	275°C
Temp Programme:	200°C for 2 minutes
	200 - 260°C @ 4°C/min
	260°C for 2 minutes

Q.8 How do you explain the results that you obtain?

You are provided with samples of tablets containing barbiturates. Extract the powdered tablet by triturating with methanol, and identify and quantify the compounds present in the methanolic solution.

5.3.7 Benzodiazepines.

Some classes of drugs contain a wide variety of structures. This is true of the benzodiazepines. Some chromatograph well, whilst others, particularly those containing hydroxide moieties, are thermally labile. This is true of drugs like temazepam, and the resulting effect is the presence of very small peaks and multiple breakdown products in the gas chromatogram (Fig 5.13.).

However, it is possible to separate some benzodiazepines, as shown in Figure 5.15., using the system described below.

Detector:	FID
Column:	BP-1 or BP-5, 0.5μm layer thickness, 25M
Split ratio:	50:1
Carrier Gas:	1ml/min
Injection temperature:	275°C
Detector temperature:	275°C
Column Temperature:	250°C

You are provided with samples thought to contain benzodiazepines. Identify and quantify any benzodiazepines present in the sample.

5.4 References

[1]   Braithwaite, A. and Smith, F. P., <u>Chromatographic methods</u>,   4th Edition,   Chapman and Hall, London, 1985.

[2]   Deutsche Ferchungsgimenshaft and the International Association of Forensic Toxicologists (1992), Gas chromatographic Retention Indices of Toxicological Relevant Substances on Packed or Capillary Columns with Dimethyl Silicone Stationary Phases.   VCH Verlagsgsellschaft mbH, D6940 Weinheim Federal Republic of Germany.

6

High performance liquid chromatography of drugs of abuse

6.1 Introduction

In addition to gas chromatography, high performance liquid chromatography (HPLC) provides another powerful method in the analysis of drugs of abuse. This method can be used to identify and quantify controlled substances, and can be used to investigate whether they form components of seizures of suspected drugs.

HPLC has a number of advantages over gas chromatographic methods, but also a number of drawbacks. The advantages are that HPLC is a non destructive method which can be used as a preparative technique if further analyses are required. Moreover, it does not require that the analyte be volatile, neither is derivatisation of the sample usually required prior to the analysis. Additionally, HPLC systems can be automated as can GC systems, for the routine screening of a large number of samples, a high throughput of samples being associated with usually short analysis times.

It is necessary for the molecule of the analyte to possess good liquid chromatographic properties, and to be detected in some way in a liquid stream. For the analysis of drugs of abuse, this is usually achieved by measurement of the absorption of ultraviolet (UV) light at a given wavelength. Further, HPLC does suffer from relatively high running costs.

The problems of poor chromatographic behaviour can be overcome by use of the correct chromatographic conditions, using one of the variety of techniques described below.

Most drug molecules absorb UV light in some part of the UV spectrum, and the amount of light absorbed will vary with both analyte concentration and chromatographic conditions. The amount of UV light absorbed can be used to achieve quantification of the analyte. This is discussed further below.

Finally, different members of different classes of drug, and indeed, different members of the same class of drug absorb different amounts of UV light of different wavelengths. This manifests itself in a UV absorbtion spectrum, which can be recorded from an HPLC eluate using a diode array detector. The information so gained can be used for drug identification.

A high performance liquid chromatograph consists of a number of component parts. These include a solvent reservoir, a pump, a sample injection system, a column for the separation process, a detecting system, and a means of recording and handling any data which has been generated.

A number of solvents (also known as mobile phases) can be used, and the solvent employed for the separation depending on the type of chromatography to be employed. For example, for relatively non-polar solutes, straight phase chromatography with organic solvents is employed, whilst for relatively polar solutes, reversed phase chromatography with aqueous solvents is used.

All solvents should be filtered and degassed prior to use. The filtration process removes any very small particles which may block the tubing or the column itself. Degassing may be achieved by either continuously purging the solvent with helium, or placing the solvent in a sonic bath prior to use. Solvents for straight phase chromatography, such as the opiates solvent described in this volume, should not be degassed using helium, or for too long in a sonic bath, because of the very volatile nature of some of the components. Degassing is desirable because at the low pressure end of the system degassing may lead to bubble accumulation in the detector, and to erroneous results.

High purity solvents are also desireable because they have much of the contamination which arises from the manufacturing process removed. However, when working at the short wavelength end of the UV spectrum (i.e below about 254 nm) the absorption of UV light by the solvent can be considerable unless high purity solvents are used. This is particularly true of some of the more frequently used solvents including acetonitrile, tetrahydrofuran, diethylether and ethyl acetate.

The solvent reservoir is connected to the solvent pump via a 5 μm filter, and a length of solvent resistant tubing. The tubing should not be soluble in the solvents employed. The pump should be capable of delivering a fixed volume of solvent in a fixed time (i.e. constant flow rate), measured in ml/min. Typical analytical rates vary from 1 - 2 ml/min, giving a pressure drop of up to 3000 psi over a 25 cm column. If solvent systems and flow rates are required which give a higher pressure than this, the packing materials become damaged, and alternative HPLC systems should be chosen.

The pump delivers solvent to the rest of the HPLC system via stainless steel tubing. The length of the tubing between the components of the system should be kept to an absolute minimum to reduce band spreading of the eluting compounds.

The injection unit is used to deliver a fixed volume of sample into the analytical system (cf. gas chromatography). A variety of sizes of sample loop are available, including 6 μl, 10 μl and 20 μl, although much larger loops are available for preparative work. When introducing the sample into the loop, the loop should be flushed with at least 10 times the volume of the sample to be injected, to ensure removal of the residual liquid from previous injections and loop washings.

The HPLC syringe differs from the GC syringe in that the open end of the needle is blunt, rather than pointed. A GC syringe is pointed because it must penetrate a septum. On no account should a pointed syringe be used on an HPLC system since this will damage the seals in the injection port.

The column type can vary both in column dimensions, and in packing materials. There are a number of variants on particle shapes, sizes, porosities and degrees of covering with bonded phases. A number of detection systems can also be employed, including U.V. detection at one or more wavelengths, diode array, fluorescence and refractive index detection. Data processing can include the use of a simple chart recorder, through to the use of a fully computerised data station.

Having introduced the sample onto the HPLC system, it is carried along by the solvent to the column, where the components of the mixture are separated. Columns are typically stainless steel tubes, 5, 10, 12.5, 15 or 25 cm long, with a variety of internal diameters. Radial compression columns are available from some manufacturers. These are contained within a pressure pack which compresses the flexible outer wall of the column onto the packing material.

The packing material contained within the column is either of silica gel (straight phase chromatography), or silica gel that has been chemically modified by bonding appropriate chemical moieties onto the silanol groups. Typical examples are C_2, C_8, C_{18}, C_{22} alkanes, the alkylphenyl and the cyanopropyl groups. However, of the modified phases, C_{18}, often referred to as ODS (from Octadecyl silyl) is the most commonly used phase. This is manufactured by the reaction of organochlorosilanes with the silanol groups on the silica gel, as illustrated in (Fig. 6.1).

Packing material particle shapes can be regular (spherical) or irregular, and come in a variety of sizes, including 3, 5 and 10 μm. Whilst the chromatographic properties of the small particles are desirable, their use generally generates an increase in operating back pressure. Perhaps more important is the increased cost of columns packed with the smaller particle sizes. For this reason, 5 μm particles are used most frequently for the analyses of drugs of abuse.

The choice of stationary phase and mobile phase is determined by the type of molecule to be separated. In both straight and reversed phase systems, the efficiency of the column is determined by the ways in which the eluant flows through the column, the so called "eddy diffusion" effect, the lateral diffusion of the molecules, and the mass transfer effects.

The eddy diffusion affects the separation because at a given solvent velocity, the molecules pass through the column via different paths, which have different lengths. This causes the molecules of each solute to spread out.

$$\bigcirc\!\!-Si-OH \quad + \quad C_{18}H_{37}SiCl_3$$

$\downarrow$ Silylation

$$\bigcirc\!\!-Si-O-\underset{\underset{Cl}{|}}{\overset{\overset{Cl}{|}}{Si}}-C_{18}H_{37} \quad + \quad HCl$$

$\downarrow$ Hydrolysis

$$\bigcirc\!\!-Si-O-Si-C_{18}H_{37} \quad + \quad 2HCl$$

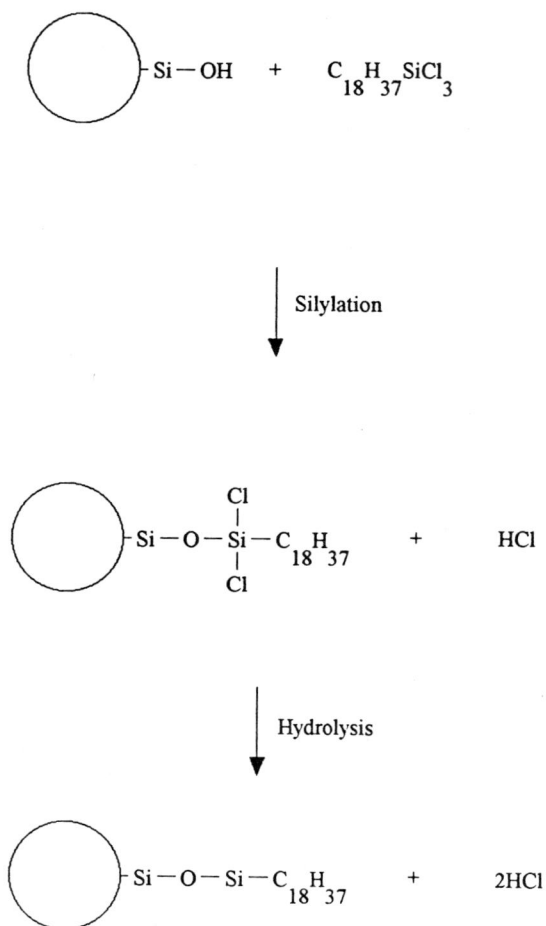

Fig 6.1 Reaction sequence used to prepare ODS packing material

This is why small, uniform, spherical particles of a given stationary phase often give better results than those of larger irregularly shaped particles of the same type.   Lateral diffusion also occurs, but to a much lesser extent in HPLC than in GC because the diffusion velocities of particles in liquids are only 10^{-4} - 10^{-5} those in the gas phase.   Finally, mass transfer effects play an important role in the separation process and in determining the efficiency of a column. It is the interaction of the solute with the stationary phase that determines the quality of the separation process. This is dealt with in greater depth in the next section.

6.2 The separation process

The separation process in HPLC is far more complex than for GC, but a correct understanding is required for good drug analysis.   The principle separation mechanisms are outlined below.

6.2.1 Normal phase chromatography

Using silica gel, for normal phase chromatography, results in a number of different kinds of interaction each of which contribute to the gel filtration process.   The relative importance of each depends upon the mobile phase chosen and the analytes in question.   They include gel filtration, partition processes (in solvents containing water), solvent exchange processes, and electronic interactions resulting from hydrogen bonding and those arising from the interaction of delocalised electrons.

Silica gel is covered in pores, some relatively large, others relatively small.   Small analytes are able to enter many of these pores and are strongly retained, whilst the larger analytes are not able to enter so many pores, and are less strongly retained, eluting more quickly.   These processes form the basis of gel filtration chromatography.

In solvents containing water, the silica gel adsorbs a layer of water onto the surface of the silanol groups.   This results in a polar layer of solvent surrounding the stationary phase, the mobile phase being relatively non polar. The analytes partition between the aqueous layer surrounding the stationary phase, and the mobile phase.   The polar analytes preferentially reside in the aqueous adsorbed layer, and as a consequence are more strongly retained than the less polar analytes.

In solvents which do not contain water, a different type of solvent stationary phase interaction occurs, known as solvent exchange.   The polar components of the mobile phase bind onto the active sites of the stationary phase.   These are displaced by the analytes in the mobile phase as they pass through the column.   As more solvent passes through the column, so the analytes are re-displaced.   However, the more lipophilic analytes are more easily displaced than the more polar analytes.   This results in the more polar analytes being retained more strongly on the column.

Electronic interactions also occur.   These include hydrogen bonds, with, for example hydroxide (Fig. 6.2a), amino moieties (Fig. 6.2b), and carbonyl groups (Fig. 6.2c).   Additionally, interactions between delocalised electrons (pi-electrons) and the lone pairs of electrons of the oxygen atoms in silanol groups result in molecules being retained (Fig. 6.3).   The stronger these interactions are, the more strongly the molecule is retained.

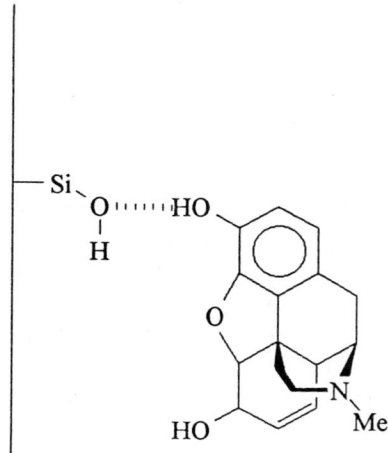

Fig. 6.2a Interaction of silanol groups on silica gel with hydroxide moieties

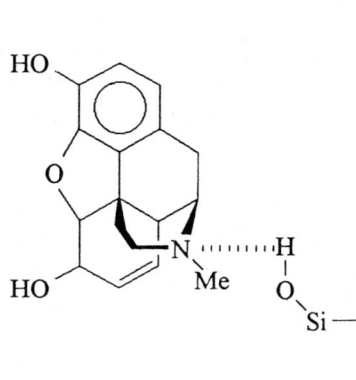

Fig. 6.2b Interaction of silanol groups on silica gel with amino moieties

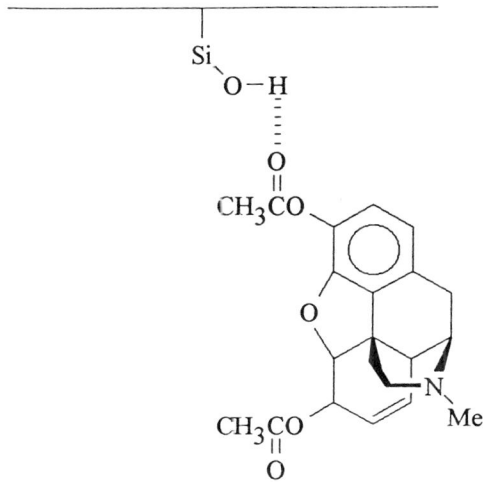

Fig. 6.2c Interaction of silanol groups on silica gel with carbonyl groups

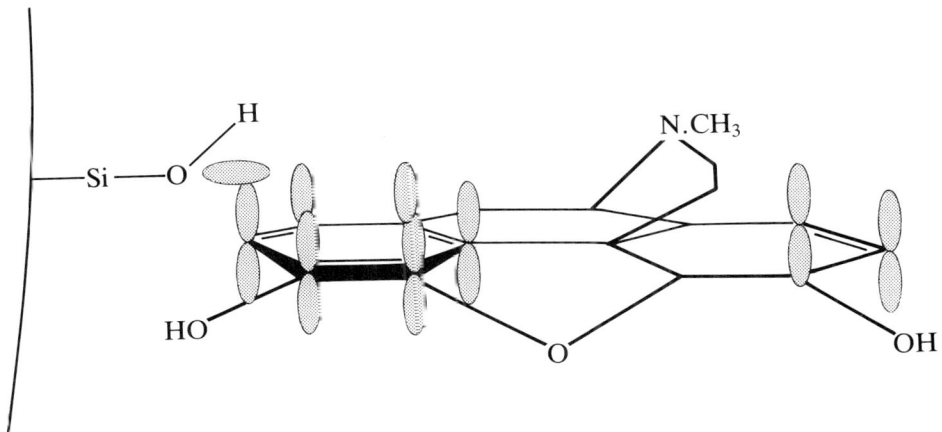

Fig. 6.3 Interaction of pi-electrons of analytes with lone pair electrons
of the oxygen atoms in silanol groups

6.2.2 Reversed phase chromatography

This form of silica gel has been modified chemically, resulting in a lipophilic stationary phase. A number of processes result in analyte separation, including gel filtration, and partition. The gel filtration operates on the same principle as for normal phase chromatography. However, the partition process is **not** the same as for normal phase chromatography,
    The organic modifier/water system in the buffer results in an equilibrium between the mobile phase and the stationary phase, since some of the organic modifier dissolves in the stationary phase. The more lipophilic solutes preferentially dissolve in the stationary phase, resulting in long retention times, whilst the converse is true of the polar analytes.

6.2.2.1 Ion suppression chromatography

Some drugs, particularly with acidic or basic moieties, acquire a charge in aqueous buffers with a neutral pH (Fig. 6.4). The charged species interact with the un-capped, free silanol groups present on the stationary phase, resulting in tailing of the analyte.
    This can be corrected by the addition of acid (for acidic drugs), or base (for basic drugs). This forces the equilibria in which the drugs participate towards the uncharged form of drug, resulting in an analyte with better chromatographic properties.

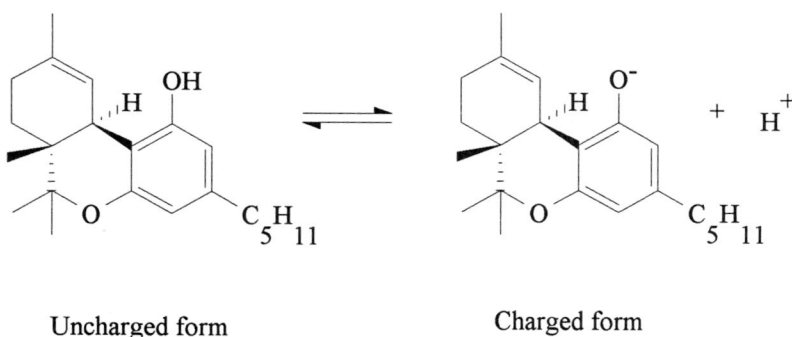

Uncharged form   Charged form

Fig. 6.4 Equilibrium between charged and uncharged form of
Δ^9-THC in aqueous solution

The acids or bases employed should be of sufficient ionic strength to buffer the highest concentration of drug examined, and the pH of the eluant should lie in the range 3 to 8. Above pH 8 the silica gel will dissolve, below pH 3 the siloxane bond begins to hydrolyse.

The whole HPLC system should always be washed with copious amounts of neutral aqueous buffer, water and then methanol after use of acidic or basic buffers, because such buffers are corrosive and are likely to precipitate out of solution, causing expensive damage to equipment.

6.2.2.2 Ion pairing systems and reversed phase chromatography

Another approach which can be adopted to alleviate the problem of charge on the solute drug is to deliberately induce a charge on the solute, and then to counter this with an ion of opposite polarity, to produce a neutral species having good chromatographic properties. The counter ions commonly used include tetramethyl-ammonium ions and n-alkylamines (eg. hexylamine) for acidic compounds and alkylsulphonates for basic drugs.

Two separation mechanisms are thought to operate. In the first, the ion pairing agent and analyte form a neutral species. The alkyl chain of the ion pair results in good chromatographic properties, and the analytes separate on the basis of partition between the stationary and mobile phases. The second mechanisms involves the ion pairing agent interacting with the stationary phase, forming a surface on which ion exchange can occur. The drugs then separate on the basis of their strengths as acids or bases. In practice, it is likely that both mechanisms may be involved.

6.2.2.3 Gradient elution systems

Some mixtures of drugs contain components which interact strongly with the stationary phase, in addition to those which interact weakly. Under such conditions it may be necessary to use a gradient of increasing eluting power to resolve all of the compounds of the mixture. This technique is better suited to reversed phase chromatography systems than to straight phase ones because the system must be equilibrated at the commencement of each chromatogram. For straight phase systems, this can take a considerable period of time.

6.3 Detector systems

Having separated the drug mixture on the HPLC column, the next step is to detect them as they elute from the column.

The most commonly employed detector is the ultra-violet (UV) detector, operated at a particular wavelength. The compound elutes from the column, and passes through a flow cell in the detector, through which light of a known wavelength passes, onto a photomultiplier or photocell. As the compound passes through the flow cell, it absorbs some of the UV light, decreasing the amount which is incident on the detector. This decreases the output from the photocell and this change is amplified and the signal recorded on a data handling device.

6.4 Quantification of samples and preparation of calibration curves

The absorbance of UV light by a drug can be used to quantify the amount of that drug present in the mixture under analysis. To perform this operation, the linearity of the absorbance response of the detector to the drug in the concentration range of interest must first be established. This assumes that absorbance obeys the Beer-Lambert relationship:

$$A = E.C.l$$

where A = the absorbance; E = the coefficient of extinction of the drug; C = the concentration of the drug, and l = the path length of the flow cell.

E and l are constant for any particular drug on any given instrument, so that the absorbance becomes directly proportional to the concentration of the drug in the column eluate. Single point estimates of concentration, line of best fit, or regression equations can all be used to determine the concentration of a sample in a solution.

When using the line of best fit, or the regression analysis method, a curve should be prepared using a range of drug concentrations which are likely to be encountered in samples of forensic interest. A plot of absorbance versus concentration of drug is made, and a line fitted, or regression equation calculated. The absorption of the compound under investigation in the mixture under analysis is noted, and from the regression equation, the amount of compound present can be calculated. Examples of these methods are illustrated below.

Whether HPLC or GC is chosen for the quantification process, a number of basic principles must first be followed before proceeding. Firstly, if for example, a drug, within a powder, is to be examined, then it should be freely soluble in the solvent chosen for the injection into the chromatographic system and should not precipitate in the HPLC eluant at the top of the column.   Methanol is frequently chosen because a wide range of different drugs dissolve in this solvent.

If a powder, resin or plant material is to be examined, it may not be totally soluble in the solvent chosen. This leads to the problem of extracting all of the drug from the seizure. This can be achieved by repeated extraction into the same solvent, either at room temperature, or if a large amount of drug is available, by soxhlet extraction, providing that heat does not cause sample decomposition, This is suitable, for example, for an etheral extraction of Cannabis resin. In such cases, the extraction should be monitored by spot tests and/or TLC to determine whether the extraction is complete and that no sample decomposition is taking place. Cold extraction is often accomplished by triturating the drug with a solvent in a mortar and pestle. The triturate is then recovered by filtration and the filtrate washed with more solvent until no more drug is extracted. This can be monitored by colour test or TLC.

In both GC and HPLC, baseline resolution of compounds must be achieved in the chromatography so that a peak height or area can be assigned to one compound alone.

Additionally, it is important that calibration curves in HPLC are prepared in the same batch of solvent in which the sample will be analysed. This is important because small differences in pH can lead to different extinction coefficients, when measuring U.V. absorption, therefore rendering the quantification process inaccurate.

When preparing calibration curves, a wide enough range of concentrations should be used to ensure that sample concentration will fall in the linear range of the curve. If, for example, a range of 0 - 1.25 mg/ml is used to establish linearity, and the dissolved sample is prepared at a concentration of 1 mg / ml, it is simple to see that the concentration of the drug will lie somewhere on the linear range of the chromatographic system. Further, when preparing the calibration curve, if this is the chosen method, then the samples should be injected starting with the lowest concentration, increasing to the highest in ascending order. This reduces the risk of column priming. Between each sample, an injection of the solvent alone should be made. This ensures that the chromatographic system is clear of any contamination which may cause inaccurate results.

6.4.1 Single point estimates.

Although this method is one of the quickest available for quantification purposes, it is also prone to a number of inaccuracies. The system assumes linearity, and that the concentration of the sample is within the linear dynamic range of the detector.

To use this method, several simple steps are followed :-

1. Obtain the peak height or area for a sample of known concentration, of the order of magnitude expected in the sample.

2. Obtain the peak height or area for the sample, having run a blank between the sample and the standard.

3. Calculate the ratio of the areas or heights.

4. Assuming the concentration in the sample to standard are in the same ratio, calculate the concentration in the sample solution.

5. Calculate the concentration in the dry sample as a percentage of the total mass.

In the example given below, a single assay estimate has been used to estimate the amount of cocaine in a seizure of leaves from the plant Erythroxylon coca, the source of cocaine.

The leaves were extracted with 75% aqueous methanol at 100 mg dry weight of leaves per ml of solvent. The extract was diluted 50 times to bring the estimated amount of cocaine into the same linear range range as the standard solution used. The HPLC system described later in this chapter (Section 6.6.4) was employed.

Cocaine standard, 0.05 mg/ml = 6100472 area counts
Leaf extract, 2 mg leaf / ml = 2670790 area counts

Therefore the leaf sample contained :-

$$\frac{2670790}{6100472} \times 0.05 \text{ mg/ml} = 0.0218 \text{ mg/ml}$$

The sample was prepared at 2mg leaf per ml solvent, so $0.0218 \mu g$ was present in $2 \mu g$ leaf, and so the leaf contained $\frac{0.0218 \mu g}{2 \mu g} \times 100\%$

= 1.1% cocaine on a dry weight basis.

So, the leaf contains 1.1 % cocaine, on a dry weight basis.

6.4.2 Two point calculation: Y = mX + c

This system still assumes that the detector response is linear in the range to be tested. A standard, prepared at a concentration lower than that expected in the sample, and one with a higher concentration, are chromatographed, and the peak height or areas obtained.

The values of peak area (Y) and concentration (X) are then inserted into 2 separate equations for a straight line (Y = mx + c). Values of m and c can be calculated by solving the simultaneous equations.

Standards

One of the equations is then re-arranged to give the concentration in terms of peak height (or area), and the value of peak height (or area) for the sample is obtained by experimentation, and inserted into the equation. The value for concentration is then calculated and from this the sample concentration as a percent of mass (for a solid sample). If a diluted liquid sample has been analysed then the drug concentration can be similarly calculated.

6.4.3 Linear regression

There are a number of ways of calculating the equation of a straight line through a number of points on a calibration curve. This means of quantification is the best of the three in that a wide range of concentrations can be examined to determine whether the range is in fact linear. The method also takes into account variation in detector response between samples. The disadvantages include the fact that a large number of standard solutions need to be injected (n>12) for the method to be statistically valid, and so the method is most suited to equipment that has autosampling facilities. An example is shown below for the calculation of the concentration of diamorphine in a case sample. The data is presented in Table 6.1.

To perform the method, the equation of a straight line is assumed to be of the form $Y = mX + C$　(1).

Therefore, for all the points,

$$\Sigma Y = nc + m\Sigma X \qquad \text{(2) and}$$
$$\Sigma XY = c\Sigma X + m\Sigma X^2 \qquad \text{(3) [eqn (1) x X].}$$

where n is the number of data points

Table 6.1 Data obtained for calibration curve of diamorphine
using the HPLC system described.

Peak Area Diamorphine (y)	Diamorphine Concentration (x)	(x^2)	xy
34286	0.075	0.01	2571
33928	0.075	0.01	2545
73809	0.156	0.02	11514
77488	0.156	0.02	12088
152515	0.313	0.10	47661
156272	0.313	0.10	48835
315243	0.625	0.39	197027
319580	0.625	0.39	199738
596227	1.250	1.56	745284
611183	1.250	1.56	763979
1094441	2.500	6.25	2736103
1156063	2.500	6.25	2890158
Total 4621035	9.84	16.66	7657656

Taking equation 2 and 3 and substituting in the summation values, as required:

$$\Sigma Y = nc + m\Sigma X$$
$$4621035 = 12c + 9.838m \qquad (4)$$
and
$$\Sigma XY = c\Sigma X + m\Sigma X^2$$
$$7657656 = 9.838\ c + 16.66m \qquad (5)$$

Solving these simultaneous equations for m and c gives m = 450166 and   c = 16049

Giving the final equation Y = 450166 X + 16049.

These figures can also be calculated on a computer spreadsheet, which shows the regression equation to be :-

$$Y = 450057\ X + 16152$$

A spreadsheet was used to generate the data for the calibration curve (Fig. 6.5), and the calculation of the amount of diamorphine and mono-acetylmorphine in the sample.

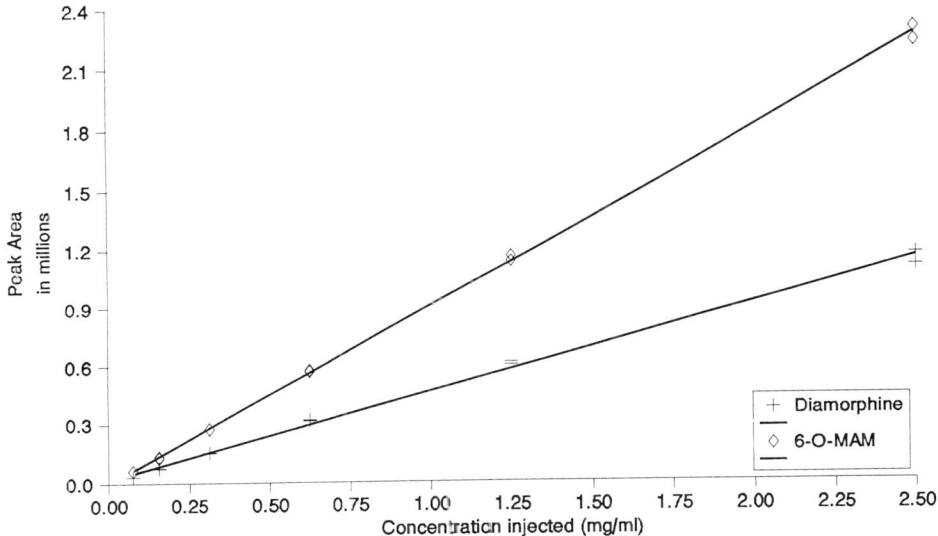

DAM : Y = 450057 X + 16152
MAM : Y = 90678 X + 2889

Fig. 6.5 Calibration curve for diamorphine and 6-0-monoacetylmorphine
using the HPLC system described

A few important points should be followed when preparing the calibration curve. Firstly, linearity should always be established, and the regression equation recalculated, if a new solvent batch or column is to be used. Secondly, the calibration curve should be prepared starting with the lowest concentration to minimise the effect of column priming. Thirdly, between each of the replicates and concentrations, a solvent blank should be chromatographed. This ensures that the injection loop and syringe are clean before the next analysis. For good reproducibility, the injection loop should be completely full. At least 12 data points, 2 at each of 6 concentrations are required for the calculation of a regression equation. Whilst peak height or peak area can be used for the final calculation, the same parameter should be used throughout. The use of peak height is preferable since this also allows column performance to be monitored.

It is important to ensure that only the linear part of the HPLC calibration curve is employed. The calibration curve shown below (Fig. 6.6) illustrates that as concentration increases, so the detector becomes saturated. Whilst it is possible to fit mathematical models to the whole dataset obtained, the quantification data obtained cannot be considered reliable.   For this reason, only the linear portion of the calibration curve should be used.

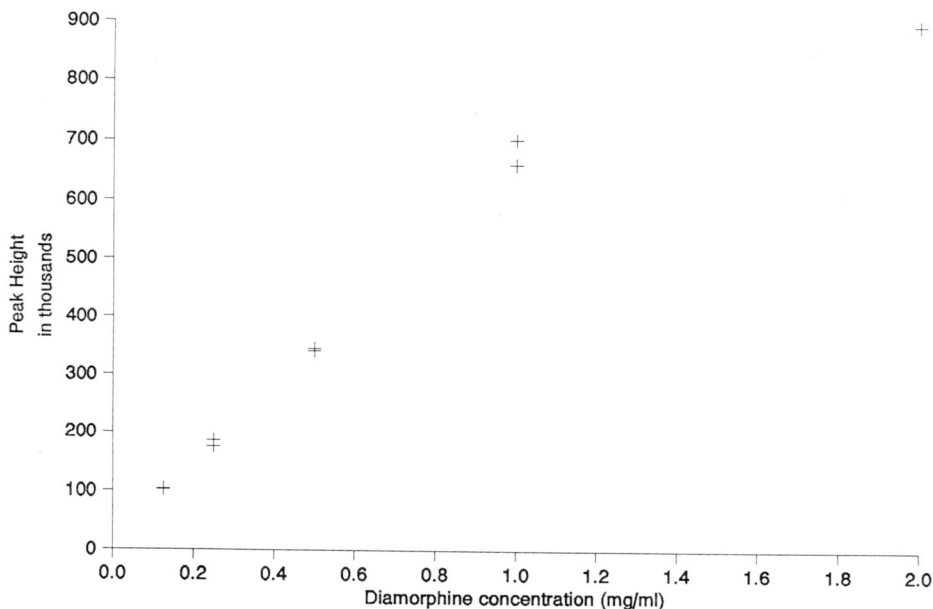

Fig. 6.6 Calibration curve for diamorphine illustrating saturation
of the detector.

6.5 Diode array detection

The use of detectors which analyse HPLC eluates at a single wavelength is limited in terms of drug identification. This is because the only data that is available from this method is a retention time. A diode array detector will scan the eluate through a number of different wavelengths from, say, 190 nm to 365 nm, to obtain a UV absorption spectrum. Thus, with this type of detector, both a retention time and a UV spectrum are obtained, making the identification of the eluate in question far more positive. This is illustrated in figures 6.7 - 6.10 for a sample of diamorphine in an illicit seizure compared to that from a control of pure diamorphine.

There are, however, some drawbacks to this technique. In particular the absorption maxima recorded are only accurate to a few nanometres. However, the use of retention time and UV absorption data vastly increase the selectivity of the analytical system since both sets of data must match those of the drug standard, allowing more definitive statements to be made about the identify of the drug.

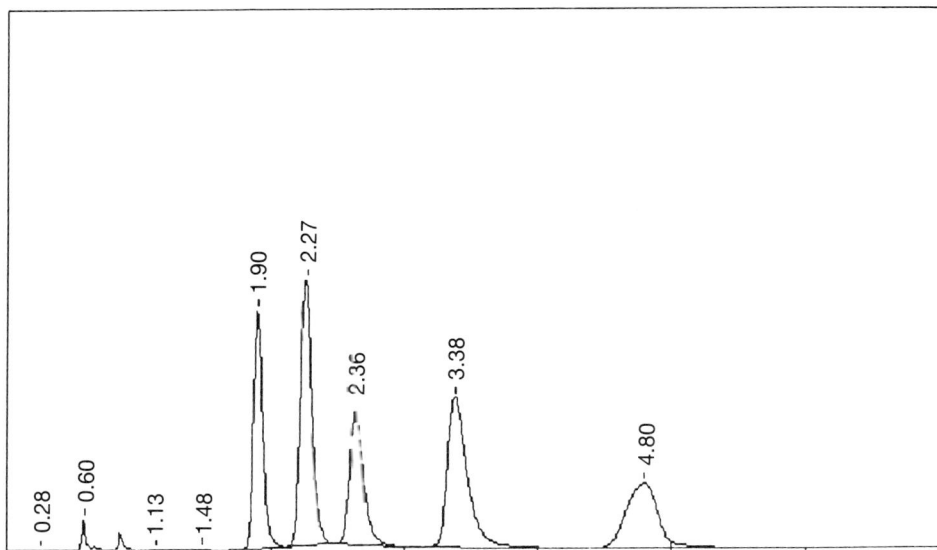

Fig. 6.7 Chromatographic separation of the opiates by HPLC (elution order;
acetylcodeine 1.40min, diamorphine 2.27min, 6-0-monoacetylmorphine 2.63min,
codeine 3.38min, morphine 4.80min)

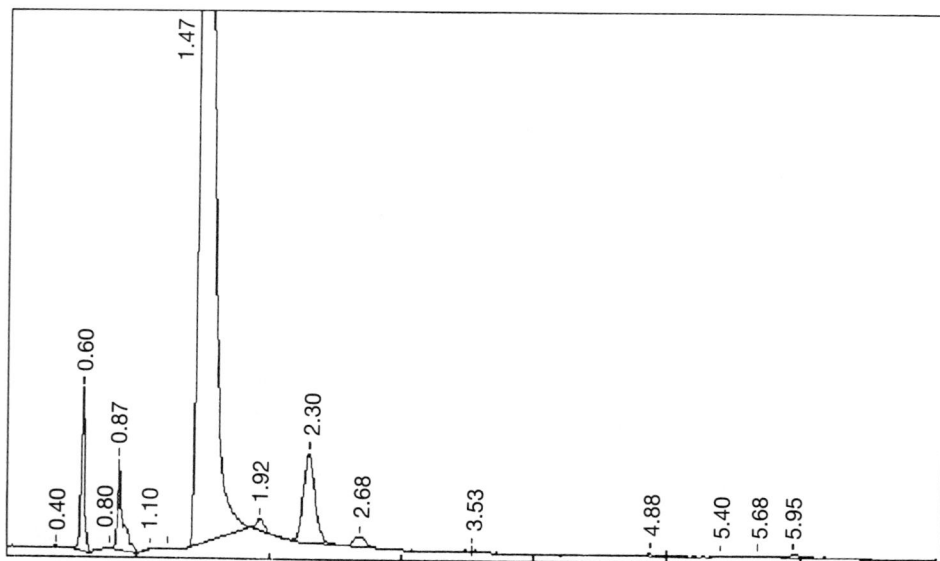

Fig. 6.8 Separation of components of a sample of a "heroin" seizure.
Elution order Caffeine (1.47min), Acetylcodeine (1.92min), diamorphine (2.30min),
6-0-monoacetylmorphine (2.68min)

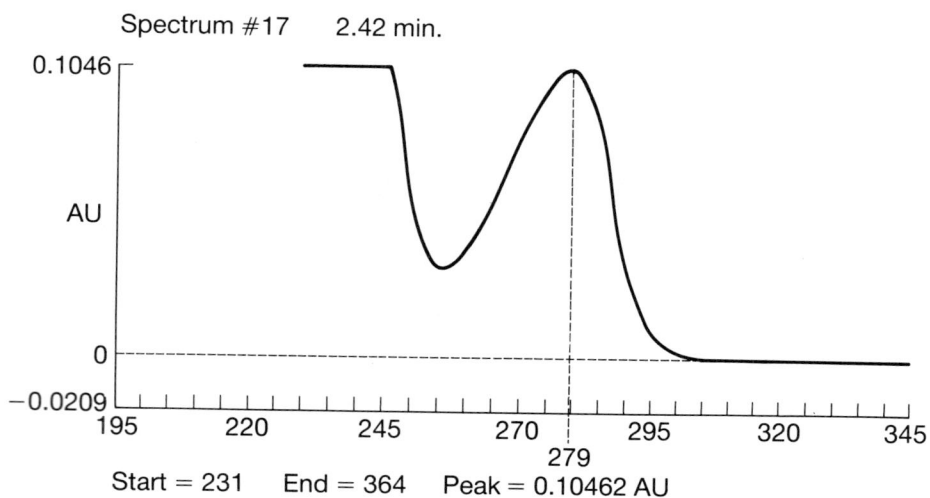

Fig. 6.9 Diode array UV spectrum of diamorphine in standard mix

Spectrum #17   2.43 min.

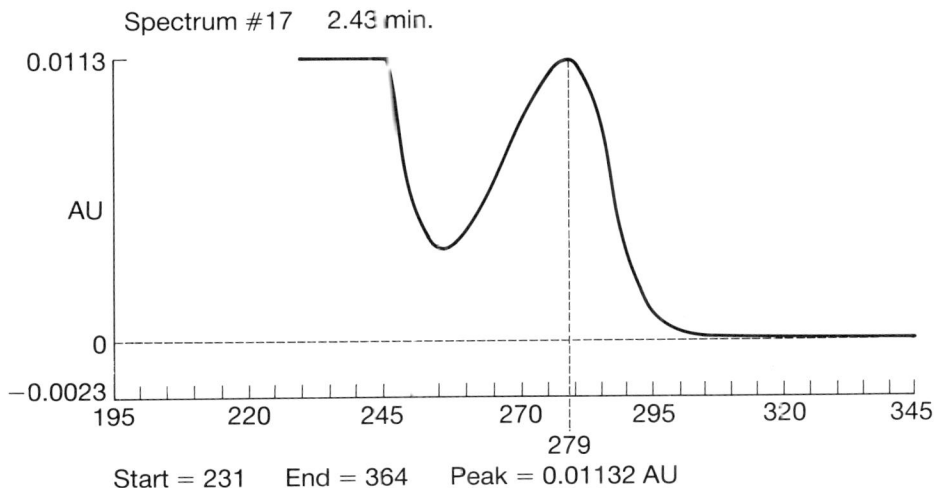

Start = 231   End = 364   Peak = 0.01132 AU

Fig. 6.10 Diode array UV spectrum of diamoprhine from "heroin" sample.

6.6 Student exercises

General methods for the analysis of abused drugs by HPLC have recently been reviewed [1].   Additional methods are described in the United Nations ST/NAR series of booklets, and Clarke's Isolation and Identification of Drugs. Below are described some systems which illustrate the theoretical points discussed above.

6.6.1 Normal phase chromatography

6.6.1.1 Analysis of opiate drugs

You are provided with samples of morphine, 6-O-monoacetylmorphine, diamorphine, codeine and acetyl codeine. Using the HPLC system described chromatograph them alone and as a mixture, at a concentration of 1 mg/ml in methanol.

Column:	Spherisorb silica gel 5 um 12.5 cm x 4.6 mm i.d.
Eluant:	iso-octane / diethyl ether / methanol / water / diethylamine / 400 / 325 / 225 / 15 / 0.65 by volume.
Flow Rate:	2 ml/min.
Injection Volume:	6 μl
Detector Setting:	280 nm

> **Q.1** How do you explain the elution order that you observe in terms of the structures of the opiates and the physico-chemical interactions which contribute to the separation process?

6.6.1.2   Quantification of diamorphine in an illicit sample of heroin

Using the opiate standards and adulterants with which you are provided, establish that (i) baseline resolution can be achieved for the opiate drugs and (ii) the detector response is linear to these drugs, in the concentration range likely to be of forensic interest.   Following the same methodology as in the example illustrated earlier, quantify the diamorphine in the mixture with which you are provided. Prepare a laboratory report on your findings and discuss possible sources of error

> **Q.2** Why is baseline resolution necessary?
>
> **Q.3** Why must linearity be established?

6.6.2 Reversed phase chromatography

6.6.2.1 Barbiturate drugs

You are provided with a series of barbiturate drugs. Using the system described obtain chromatographic data for each of them.

Column:	Spherisorb ODS-2 5 μm.
	25 cm x 4.6 mm i.d.
Eluant:	40 % acetonitrile in water.
Flow Rate:	1 ml / minute
Injection Volume:	6 μl
Detection:	235 nm 0 - 1 A.U.

Q.4 You are provided with some data obtained previously (Fig. 10.4). How do your results compare with this data and how do you explain this?

Q.5 How do you explain the elution order that you obtain?

6.6.2.2 Reversed phase chromatography and ion suppression - cannabinoids

Using the standards provided and the system described below, obtain chromatographic data for the cannabinoids.

Column:	Spherisorb 5-ODS. 5 μm.
	25 cm x 4.6 mm i.d.
Solvent:	Methanol / Water / Acetic Acid 85 / 14.2 / 0.8
	by volume
Flow Rate:	1.5 ml/min
Detection:	230 nm 0 - 1 A.U.

Q.6 How do you explain the elution order that you observe?

Q.7 Why is acetic acid included in the solvent system?

6.6.2.3 Reversed phase chromatography - diode array detection

Diode array detection can be used for the identification of drugs in a mixture. A model system which can be used to illustrate this is that used for the HPLC of cocaine and metabolites.

Using the system described by chromatograph cocaine and it's breakdown products.

Column:	Spherisorb 5 ODS-2, 5 μm
	12.5 cm x 4.6 mm i.d.
Eluant:	Methanol / Water / 1 % Phosphoric acid /
	hexylamine 35/65/100/1.4 by volume
Injection Volume:	6 μl.
Detection:	230 nm 0 - 1 A.U.

You are provided with a sample thought to contain cocaine.   Identify and quantify any cocaine present.

6.7 References

[1]   Caddy, B. "The use of high performance liquid chromatography for the detection and quantification of abused drugs", in The Analysis of Drugs of Abuse. T.A. Gough (ed). John Wiley and Sons, 1991.

7

Ultra violet spectroscopy of drugs of abuse

7.1 Introduction

In addition to chromatographic methods for the analysis of drugs of abuse, spectroscopic methods can be employed. These include the use of ultra violet spectroscopy, a technique which is relatively cheap, quick and simple. The data so generated can be used to both identify and quantify a drug of abuse.

Ultra violet spectroscopy relies on the absorption of light energy, in the wavelength range 190 - 350nm. Light in this region of the electromagnetic spectrum raises the energy levels of the electrons within a molecule, from the lowest energy level, the so called "ground state", to higher energy levels. Each transition requires a given amount of energy, provided by light of a particular wavelength, and it is the electronic structure of the whole molecule that determines the characteristic ultra violet absorption spectrum.

Contributions to the spectrum are made by electrons of different types. The electrons associated with a normal isolated single bond, termed sigma (σ) electrons can be excited by high energy light, at below 200nm, to what is called an antibonding orbital. The energy level of one of the electrons in the bond is raised, whilst the other remains involved in the bonding process. This region of the UV spectrum is of little use to the drug chemist. The electrons associated with double and triple bonds are termed pi electrons, and require less energy to raise them from the ground state to an antibonding orbital, and so absorptions are observed at higher wavelengths. Similarly, lone pair electrons found on oxygen atoms in hydroxide moieties, and on nitrogen atoms of amino groups are also raised to the antibonding pi state when irradiated with light of a suitable wavelength.

A double bond responsible for the absorption of UV light is known as a chromophore. The wavelength at which absorption occurs can be affected by a number of structural features within the molecule being analysed. If two or more double bonds are joined by single bonds, they are said to be conjugated. The effect of conjugation is to delocalise the electrons, and hence the amount of energy required to raise them from the ground state to the antibonding state is reduced.

In addition to conjugation, the presence of lone pair electrons on a functional group, joined to the chromophore by a saturated bond, may also affect the absorption maximum.   Examples of this include the lone pair electrons found on amino moieties.   The lone pair may be co-ordinately unsaturated (as in the free base of a drug), or co-ordinately saturated, as in the hydrochloride salt.   The effect on the UV spectrum of the saturated system is less than that of the unsaturated system.

When an absorption peak, obtained under one set of conditions moves from a given wavelength to another, under a different set of conditions, a "shift" in the absorption maxima is said to have occurred.   If the shift is to a higher wavelength, it is known as a bathochromic shift, whilst if it is to a lower wavelength, a hypsochromic shift is said to have occurred.   Such shifts can be of value in drug identification (section 7.3).   Additionally, a shift may be associated with an increase, at the maximum.   If the amount of absorption increases, a hyperchromic shift is said to have occurred, whilst if there is a decrease in absorption, a hypochromic shift is said to have resulted.   Again, such effects can be employed in drug identification (section 7.3).

7.2 Recording of an ultra violet absorption spectrum

The sample should be freely soluble in spectroscopic grade solvent. Additionally, modifiers of pH should be as pure as possible, because of the problem of absorption at the low wavelength end of the spectrum.   For example, spectra in sodium hydroxide, used to increase the pH of the solution, cannot be recorded at wavelengths below 225nm.   Two "matched" cuvettes are usually required to obtain a spectrum on modern instruments.   The cells should be clean, free from contamination, and made from silica glass if the full UV spectrum is to be recorded.

The sample is placed in one cuvette, the solvent alone in the other, and the samples placed in the UV spectrophotometer.   The control, reference cuvette, is usually placed in the reference beam of double beamed instruments. The absorption spectrum, and the wavelength of any maxima is recorded. The absorption maxima of the sample can then be compared to tabulated values of wavelength of maximum absorption obtained for control substances recorded under identical conditions.

7.3 Drug identification

Whilst it is not possible to definitively identify a particular drug on the basis of its ultra violet spectrum alone, a number of features of the UV spectrum can be employed to aid in sample identification.   The first of these is comparison of the shape of the spectrum, and the wavelengths at which the absorbance maxima occur, with tabulated values for controls substances, prepared in the same solvent.   It is important to use the same solvent since spectra do vary if different solvents are employed.

Alteration of the pH of the solvent can also be used.   For example, morphine undergoes a large bathochromic shift in sodium hydroxide in methanol, when compared to the spectrum obtained in hydrochloric acid in methanol alone (Fig. 7.1).

The point where these two spectra cross, referred to as the isobestic point, demonstrates the same absorbance value no matter what the pH.   The wavelength and absorption value at the isobestic point is characteristic of morphine but not necessarily unique to this compound.

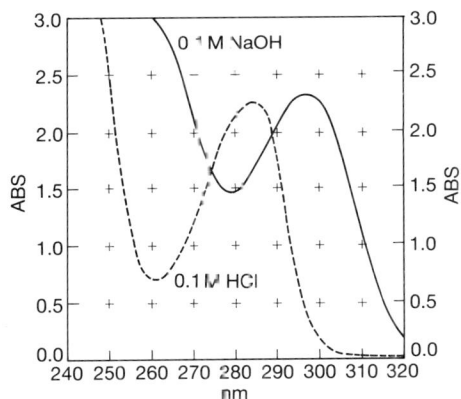

Fig. 7.1 Ultra violet spectrum of morphine in (i) 0.1M sodium hydroxide
in methanol, and (ii) 0.1M hydrochloric acid in methanol

Additionally, hyperchromic and hypochromic changes can be used in drug identification.

This same approach can be employed for the characterisation of barbiturates.   Their absorption is measured at 249nm and 260nm in each of 0.45M sodium hydroxide and borate buffer, pH 10 solutions.   The difference in absorption at 249nm for the solutions at the two pH values is recorded and the same measurement is made at 260nm: a numeral value, characteristic of the barbiturate, is then obtained by ratioing their absorbance differences.

An example of this is provided below, using phenobarbitone (Fig. 7.2). The absorbance ratios are calculated and compared to tabulated values. Examples include amylobarbitone +0.12 pentobarbitone -0.05, phenobarbitone -0.09, secobarbitone -0.13.   These can then lead to a partial drug identification, which requires to be confirmed using other methods.

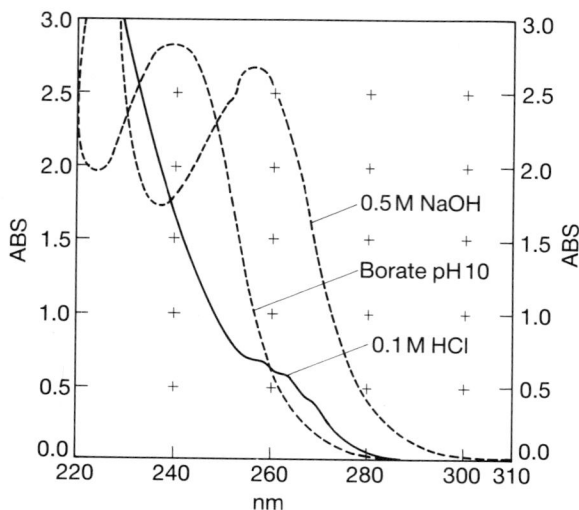

Fig. 7.2 Ultra violet spectrum of phenobarbitone obtained
under different solvent conditions

7.4 Drug quantification

UV spectroscopy can be used to quantify drugs of abuse, once linearity of
absorbance with increasing drug concentration has been established.   To
achieve this, a series of absorbance readings at different drug concentrations
should be obtained, and the regression equation calculated.   Once this has
been achieved for drug concentrations of interest, the sample can then be
compared to a point estimate.

The control drug is dissolved at a known concentration, in a suitable
solvent.   The sample is dissolved at a known concentration, in the same
solvent.   The absorbance of the control solution, and the drug are obtained,
and from the concentration of the control, the concentration of the sample is
calculated:

$$\text{Conc sample} = \frac{\text{Conc. control x absorbance of sample}}{\text{absorbance of control}}$$

7.5 Student exercises

You are provided with a series of controlled substances and their adulterants. Obtain ultra violet absorption spectra for each under acidic, neutral and alkaline conditions.

> **Q.1** Are you able to establish any ground-rules that can be used for sample identification?

You are also provided with a series of barbiturates.   Using different buffer solutions, establish absorbance ratios.   Using this data, identify and quantify any barbiturates in the material with which you are provided.

8

Infra red spectroscopy of drugs of abuse

8.1 Introduction

Once isolated from a mixture of drugs of abuse and adulterants, the drug in question must be identified. Several chromatographic means are possible, including gas chromatography and high performance liquid chromatography. However, some legislation requires that at least two independent methods of identification are employed. Infra red spectroscopy is a useful technique for the identification of drugs of abuse because it does not depend upon a separation process for sample identification. It does, however, have the disadvantage that the sample must be relatively pure if definitive spectra are to be obtained.

The infra red region of the electromagnetic spectrum is the region that causes bonds within organic molecules to bend, vibrate, and undergo other spatial distortions. Quantum mechanics dictates that these changes require discrete amounts of energy for bond vibration or distortion. These distortions or vibrations result in changes in electron density resulting in the production of a dipole moment of the bond. Thus bonds of a certain type absorb at characteristic wavelengths. These are measured at certain wavenumbers (the reciprocal of wavelength). The range of absorptions, diagnostic to certain types of bond, occur in the region 4000 - 1500 cm^{-1}. Typical values for absorbtion maxima are given below (Table 8.1). The reader should consult further texts (e.g. Williams and Fleming, 1989) for full lists of correlations.

In addition, whole molecules will vibrate when supplied with low energy infra red light over the spectral range 1500 - 750 cm^{-1}. Such vibrations are characteristic of an individual molecule. For this reason that region of the spectrum is often referred to as the fingerprint region, which is used to identify particular molecules, especially when used in conjunction with other methods, for example, gas chromatography.

**Table 8.1 Examples of regions of the infra red spectrum
where some characteristic absorbtions occur**

Group	Wavenumber (cm^{-1})
O-H Alcohol/Phenol	3500 - 3200
Carboxylic acid	3000 - 2500
C-H Alkyl	2850 - 2950
Aromatic	~3030
C=O Ketone	1730 - 1650
Carboxylic Acid	1780 - 1710
Esters	1750 - 1735
N-H Amine	3500 - 3300 (Stretch)
	1500 - 2000 (bend)
C=C	1680 - 1500

The consequence of these features is that the fingerprint region can be used to identify particular molecules, including drugs of abuse, especially when used in conjunction with other methods.

8.2 Sample preparation

The infra red spectrum of any given compound can be obtained in a number of different ways, depending upon the physical form of the molecule. These include the use of :-

1. The gas phase

2. Solution

3. Nujol mulls

4. Thin films

5. The solid phase.

8.2.1 The gas phase

Drugs of abuse generally have very low vapour pressures at room temperature, and measurement of the infra red spectrum at room temperature is of little practical importance to the forensic scientist. However, it is possible to connect infra red spectrometers to gas chromatographs, and this technique would provide a powerful tool in drug identification.

8.2.2 Solutions of samples

The measurement of the infra red spectrum can be made in chlorinated solvents which are alcohol free. However, these solvents absorb in the infra red region, including the fingerprint region, thus reducing their value in the identification of drugs of abuse.

8.2.3 Nujol mulls

The sample can be ground to a fine paste in a liquid hydrocarbon such as nujol. Whilst this technique has advantages over the previously described methods, it does suffer from the disadvantage that a large amount of sample is required. If a large sample is not available to the scientist, then this is not the method of choice. However, in the case of large seizures where a large amount of sample is available, then this method may be employed.

8.2.4 The solid phase

Infra red spectra of samples in the solid phase can be obtained from samples ground in a small amount of desiccated potassium chloride. This salt is used because it is transparent to infra red light. The sample is ground with 5 - 10 times it's weight of potassium chloride with an agate pestle and mortar. The sample is then made into a disc with a special press. The infra red spectrum of the sample within the KCl disc is then obtained.

Alternatively, if a solution of the drug is available in a dry, volatile solvent, the solution can be added directly to some ground KCl. The sample is then further mixed into the KCl by further grinding, until all traces of solvent have evaporated and the KCl disc and infra red spectrum obtained in the normal way.

8.2.5 The thin film

A small amount of the drug is dissolved in a volatile, water free solvent. A small amount of this solution is then placed on a potassium chloride disc and the solvent allowed to evaporate. The evaporation may be assisted by use of a nitrogen stream or an infra red lamp. The spectrum of the compound as a thin film of oil is then recorded. The solvent that is used should be dry, and redistilled, to remove traces of any impurities which may be present, and would be concentrated by the evaporation process.

8.3 The dispersive infra red spectrometer

There are a number of types of infra red spectrometer on the market. However, in this introduction to the technique, only the dispersive infra red spectrometer will be considered.

This instrument consists of an infra red light source, emitting across the range of wavenumbers of interest (400 - 4000 cm^{-1}). The machine can be either a single, or a double beam instrument. The double beam instrument allows compensation to be made for the absorption of Infra red light by some solvents. Using a double beam instrument, the light beam is split into two, with one half passing through the reference cell, the other through the sample. The spectrum is obtained by comparison of the absorption of the Infra red light of the sample relative to the absorption of the reference cell.

Before use, the instrument should be correctly calibrated. This can be performed using a styrene standard. This ensures that the instrument is reading the absorptions at the correct wavelengths. Deviations from precise calibrations are higher at the high wavenumber end of the spectrum (+/- 5 cm^{-1} at 5000 cm^{-1}) than at the lower wavenumber end of the spectrum (+/- 1 cm^{-1} at wavenumbers < 2000 cm^{-1}). Accuracy is especially important when the principle peak method (i.e. identification of the compound through the wavenumbers at which the greatest absorptions occur) is employed. An infra red spectrum of styrene, and the 10 principle peaks are given (Fig. 8.1).

8.4 Spectrum interpretation

Having prepared the sample, and obtained the infra red spectrum, the absorptions observed should then be interpreted. These absorptions will be due to a number of vibration types within the molecules, occurring at characteristic wavelengths depending on the type of bond, and vibrational transitions.

8.4.1 The hydroxyl group

Absorptions for the stretching of the hydroxide group occur around at 3600 cm^{-1}. However, both intermolecular and intramolecular hydrogen bonding will both affect the wavelength at which the absorptions occur, and also the characteristics of the peak shape (Table 8.2) below.

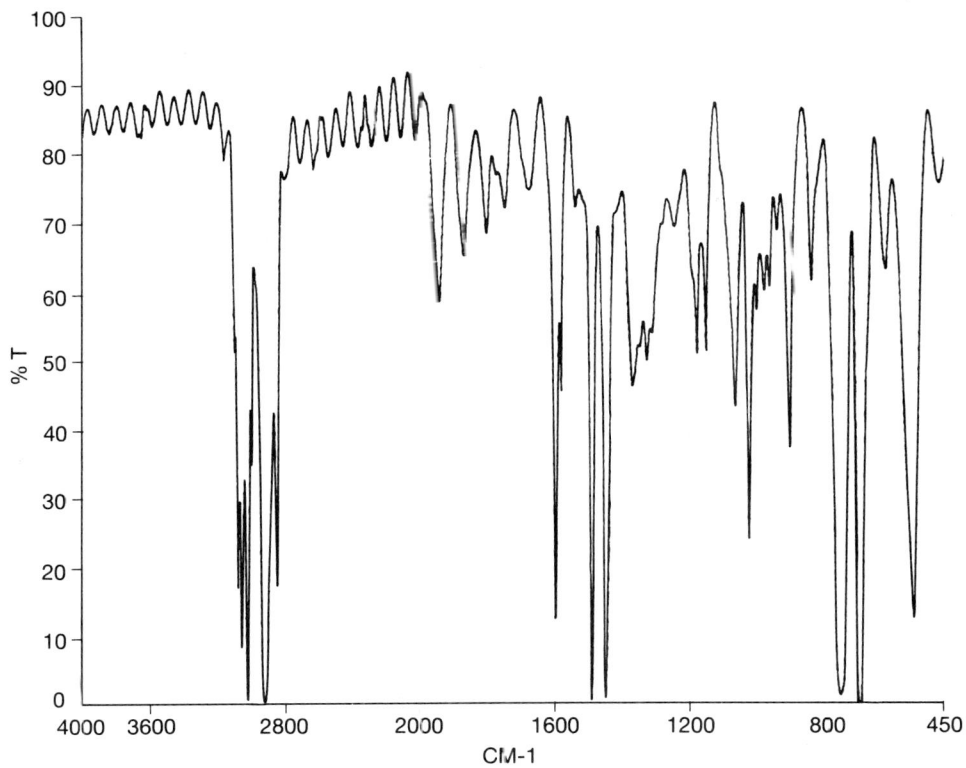

Fig. 8.1 Infra red spectrum of styrene

Table 8.2 Characteristic absorbtion maxima for hydroxide groups in the infra red spectrum

Hydroxyl Group	Wavenumber	Peak Shape
No H bonding	3600	Sharp
Intramolecular H bonding	3400	broad
Intermolecular H bonding	3000	very broad

8.4.2 Amino, imine and amide groups

Amino groups show characteristic absorptions at 3300 - 3500 cm^{-1}. N-H bending of primary amines may also cause absorptions in the region 1560 - 1650 cm^{-1}, although these absorptions may be very weak. Corresponding absorptions at 1490 - 1580 cm^{-1} are observed for secondary amines, although again these may be very weak.

8.4.3 The C-H stretches and deformations

C-H stretching usually occurs with corresponding absorptions in the region 2850 - 2960 cm^{-1}. Deformations of this type of bond, e.g. bending and twisting, occur with absorptions in the region 1430 - 1470 cm^{-1}. Alkene and aryl C-H stretching may cause a small, but sharp and distinctive peak at 3075 - 3095 cm^{-1}, although this may be masked behind other absorptions arising from other functional groups, especially if a hydroxyl moiety is present.

Out of plane deformations by C-H groups cause characteristic absorptions in the region 1430 - 1470 cm^{-1}. However, interpretation of these is difficult because these absorptions occur in the high frequency end of the fingerprint region.

8.4.4 Double bond stretching

Alkene C=C stretches are observed in the 1620 - 1680 cm^{-1} region of the spectrum. Aromatic C=C stretches may be observed at about 1500, 1580, and 1600 cm^{-1}.

8.4.5 The carbonyl group

This absorption is frequently the strongest of the whole infra red spectrum, and is centred at 1715 cm^{-1}. The frequency at which the absorption occurs is influenced strongly by the substitutions of the carbon atom in this bond. If substituted with an electronegative moiety (for example a chlorine atom), the removal of electrons from the carbonyl bond decreases the bond length, and more energy is required to make the bond vibrate. Thus such absorptions due to the vibrational transition occur at wavenumbers in excess of 1715 cm^{-1}. The converse is true for those atoms which donate electrons or have lone pairs present (e.g. a nitrogen atom). The net effect of different substituents of the carbon atom on the observed frequency at which absorption occurs is additive.

8.4.6 The fingerprint region

This region is difficult to interpret, but, taken as a whole is useful for sample identification. The difficulty in interpretation arises from the fact that the region contains absorptions associated with both changes in vibrations of individual bonds and those of the whole molecule. It is the fingerprint region which is the unique identifier of that molecule, and which can be used to identify a drug.

8.5 Interpretation of infra red spectra of some common drugs of abuse

Below are presented the infra red spectra of samples of common drugs of abuse, along with details of the interpretation of these, and information concerning the principle peaks. The samples were prepared in the solid state as KCl discs, with a resolution of 4 cm^{-1}. The infra red spectra and principle peaks are given at the end of this section.

It should be remembered that there will always be some variation between instruments in the absolute wavenumber value at which maximum absorbtion is observed. For this reason, the general form of the infra red spectrum should always be considered. Further, a standard spectrum of the drug in question, prepared on the same instrument as that on which the sample was analysed, should always be used for the comparison, rather than a direct comparison with a spectrum from a library from another source.

The above considerations are especially true if the principle peak method of compound identification is to be employed. Using this method, the strongest absorbtions are noted, and compared to tabulated values.

8.5.1 The opiate drugs

8.5.1.1　　Morphine

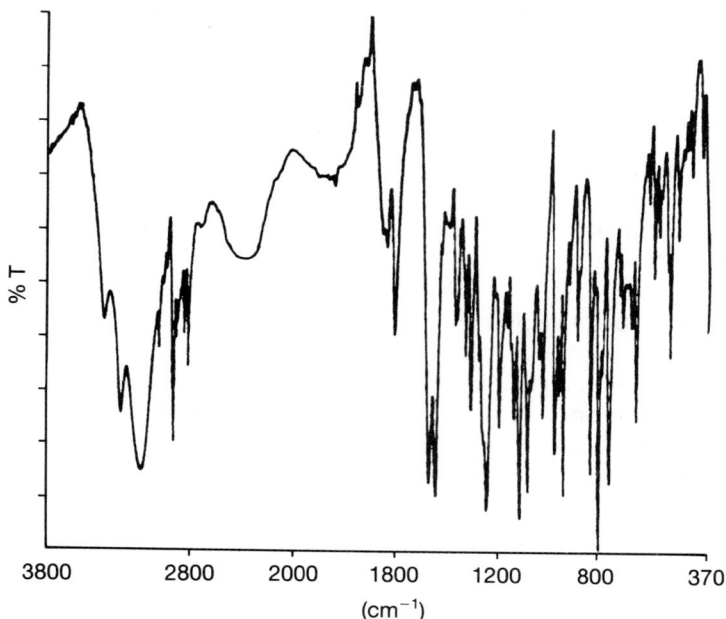

Fig. 8.2　Infra red spectrum of morphine

The peaks at 3477, 3347 and 3187 cm^{-1} are due to absorptions of the two hydroxyl groups found on the molecule. Some hydrogen bonding, both intramolecular and inter-molecular, causes peak broadening. The sharp peak at 3048 cm^{-1} is possibly due to an aryl or olefinic C-H stretch. C-H stretches also cause absorptions at 2989, 2938 and 2916 cm^{-1}. The absorption at 2822 cm^{-1} is possibly due to the stretching of the >N-Me bond whilst the absorption observed at 1634 cm^{-1} is probably due to an olefinic C=C stretch. The peak at 1604 cm^{-1} may be due to the aryl C=C stretch.

8.5.1.2　　6-O-Monoacetylmorphine

The absorptions at 3571 and 3402 cm^{-1} are due to the stretching of the O-H bond at the 3 position. The complex of absorptions centred around 2961 cm^{-1} are due to the C-H stretching, with the absorption at 2843 cm^{-1} due to the >N-Me stretch. An additional absorption at 1737 cm^{-1} is observed. This is due to the presence of the acetoxy moiety at the 6 position, and the associated carbonyl stretch. The peak at 1642 cm^{-1} is probably due to the olefinic C=C stretch, whilst the absorption at 1611 cm^{-1} may be due to the aryl C=C stretch.

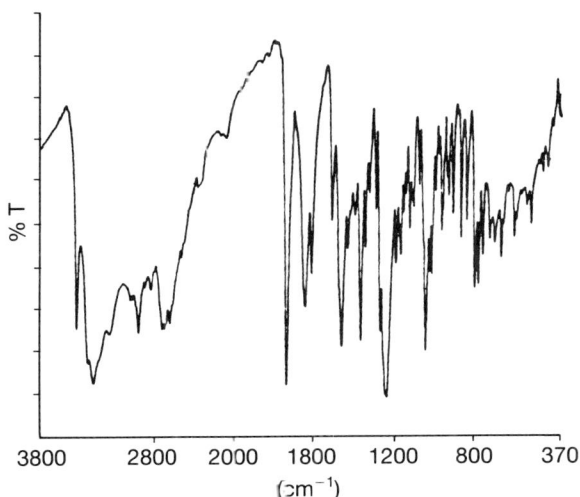

Fig. 8.3 Infra red spectrum of 6-0-monoacetylmorphine

8.5.1.3   Diamorphine

Since both hydroxide groups (cf. morphine) are now substituted, the only hydroxide peak is due to any water of hydration that may be present. An aryl C-H stretch is observed at 3038 cm^{-1}. Absorptions at 2973, 2924, 2890 and 2864 cm^{-1} are due to alkane C-H stretches, and the peak at 2802 cm^{-1} may be associated with N-Me stretch.

The carbonyl absorptions at 1762 and 1741 cm^{-1} are both visible, with the former being assigned to the C3 acetoxy substitution, and the latter assigned to the substituent at the C6 position.

The C-H bonds of the acetoxy groups would be expected to absorb strongly in the region 1365 - 1385 cm^{-1} and 1355 - 1360 cm^{-1}. Strong absorptions are indeed observed at 1350 and 1368 cm^{-1}.

The absorption at 1612 cm^{-1} is probably due to the olefinic C=C stretch.

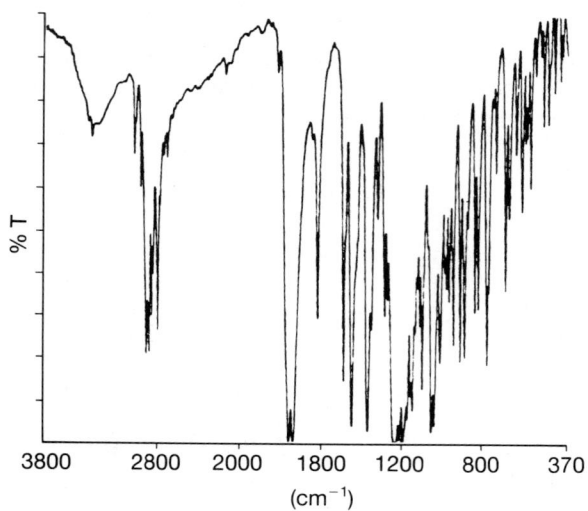

Fig. 8.4 Infra red spectrum of diamorphine

8.5.1.4 Codeine

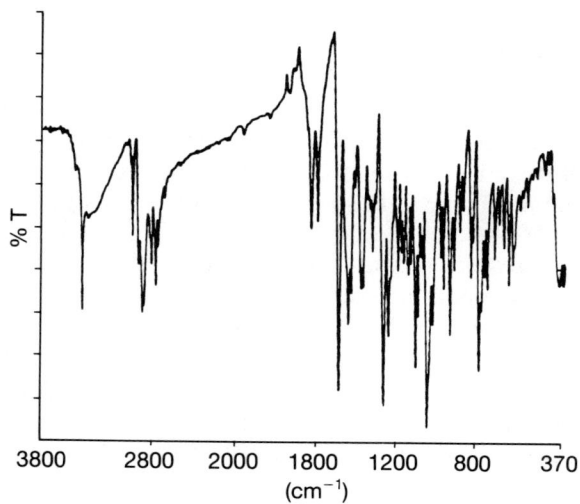

Fig. 8.5 Infra red spectrum of codeine

A sharp peak due to absorption by the C6 hydroxyl moiety is observed at 3522 cm^{-1} as is the aryl C-H stretch at 3020 cm^{-1} and the alkane C-H stretchings at 2961, 2927 and 2911 cm^{-1}. The peaks centred at 2833 cm^{-1} are probably due to the C-H stretch of the methyl ether at the C3 position.

8.5.1.5   Acetylcodeine

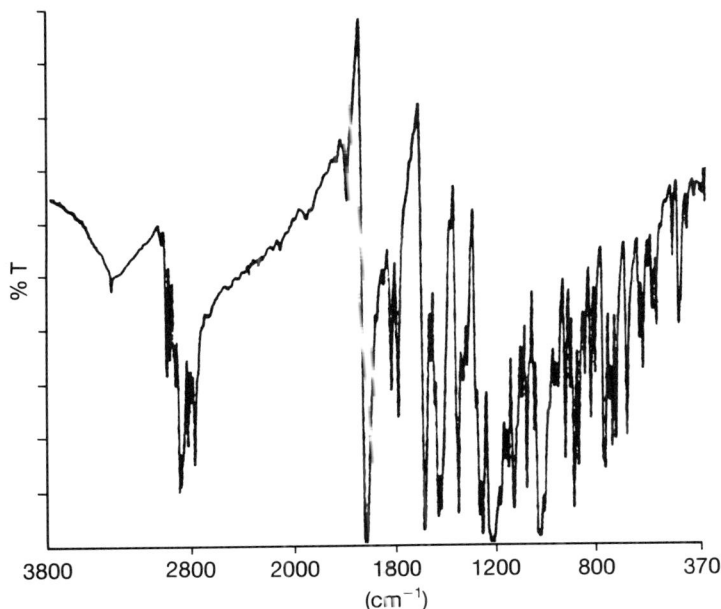

Fig. 8.6 Infra red spectrum of acetyl codeine

The infra red spectrum of this compound exhibits absorptions due to aryl C-H stretching, at 3023 cm^{-1}, and also due to water of hydration at 3450 cm^{-1}. There are a number of absorptions at 2999, 2960, 2931, 2915, 2862, and 2833 cm^{-1} which are due to alkane C-H stretching. A strong carbonyl absorption is observed at 1734 cm^{-1} as would be expected.

8.5.2 Methaqualone

The infra red spectrum of this compound in the free base form exhibits many absorptions of diagnostic value. The absorption at 3068 cm^{-1} is probably due to the aryl C-H stretch occuring, whilst those at 3010, 2947 and 2911 cm^{-1} are due to alkane C-H stretching.

     The absorption at 1677 cm^{-1} is probably due to the amide group. The free lone pair of electrons in the nitrogen atom interacting with the carbonyl group which results in an absorption at a lower frequency than would ordinarily be expected for a carbonyl group. This peak may not be fully resolved from the absorption due to the stretching of the C=N bond.

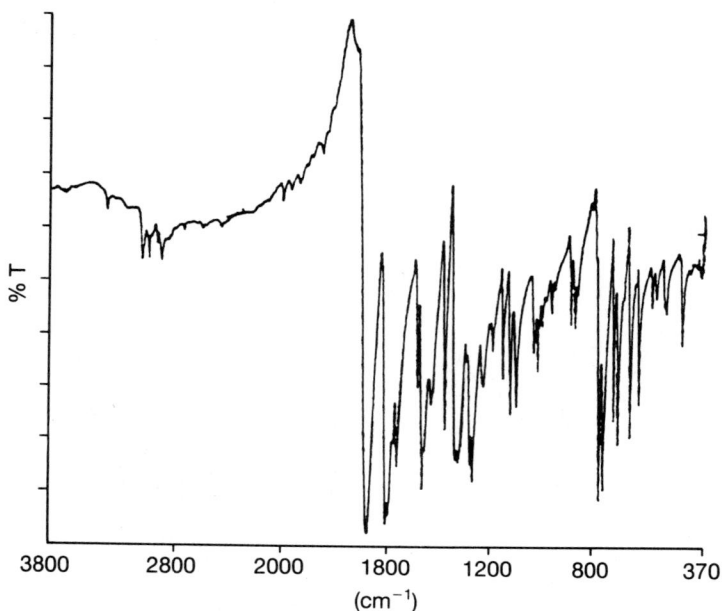

Fig. 8.7 Infra red spectrum of methaqualone

Absorptions at 1608, 1599 and 1567 cm^{-1} are probably due to the C=C stretching which is occuring in the aryl rings.

Examination of the tabulated data indicate that the 6 strongest absorptions are good diagnostic characteristics for this compound. In addition, the shape and form of the absorptions in the 3000 cm^{-1} region are also good indicators of the presence of this compound when taken into consideration with the other absorptions.

The hydrochloride salt of methaqualone also produces a characteristic spectrum. The peak at 3042 cm^{-1} is due to the aryl C-H stretch, while the absorptions at 2977 and 2989 cm^{-1} are due to the alkane C-H bond absorbing at this wavelength on stretching.

The absorption at 1724 cm^{-1} is possibly due to the carbonyl group, which may now be resolved, in the salt, from the C=N stretch, which is possibly causing an absorption at 1650 cm^{-1}. The absorptions at 1580 and 1541 cm^{-1} are due to aryl C=C stretching.

8.5.3 Cocaine

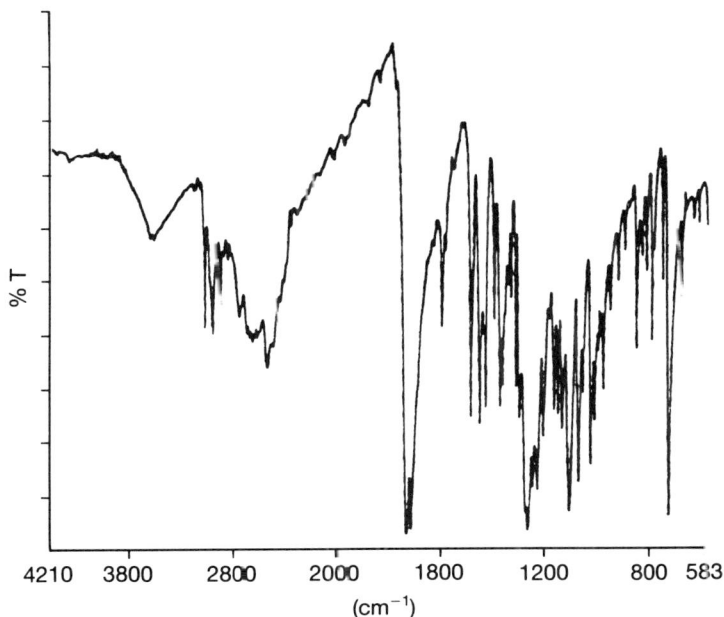

Fig. 8.8 Infra red spectrum of cocaine hydrochloride

The peak at 3022 cm^{-1} is due to an absorption arising from an an aryl C-H stretch, whilst the peaks at 2968, 2958 and 2899 cm^{-1} can be assigned to absorptions associated with the alkane C-H stretchings. The peaks at 1729 and 1713 cm^{-1} are due to the carbonyl groups found on the ester moieties, and absorptions at 1597 and 1582 cm^{-1} are due to the aryl C=C stretch. Additionally, the absorption at 2843 cm^{-1} may be associated with the C-H stretching of the methyl ester.

The five most intense peaks provide a means of identification of this compound. A comparison is made between those data obtained from this study and those from the published literature.

8.5.4 The barbiturates

The principle peaks of the infra red spectra of the barbiturates, exemplified by barbitone (Fig. 8.9), phenobarbitone (Fig. 8.10) and cyclobarbitone (Fig. 8.11), are all very closely related.

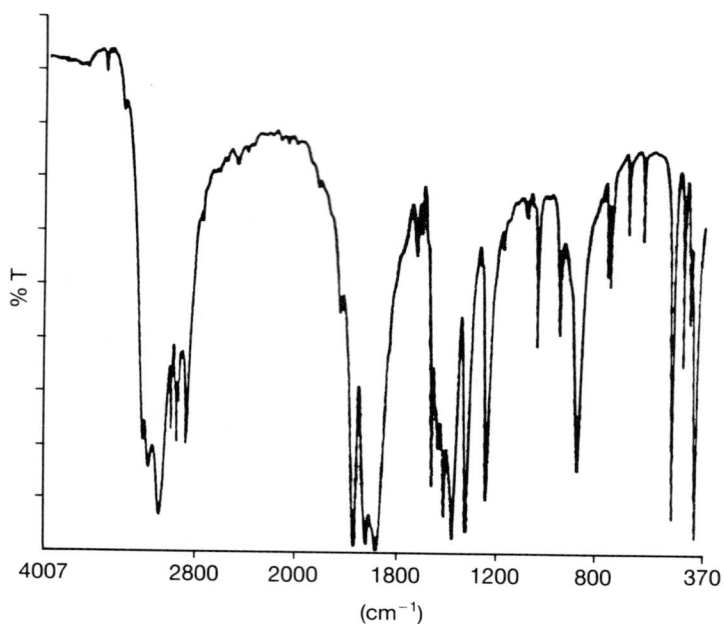

Fig. 8.9 Infra red spectrum of barbitone

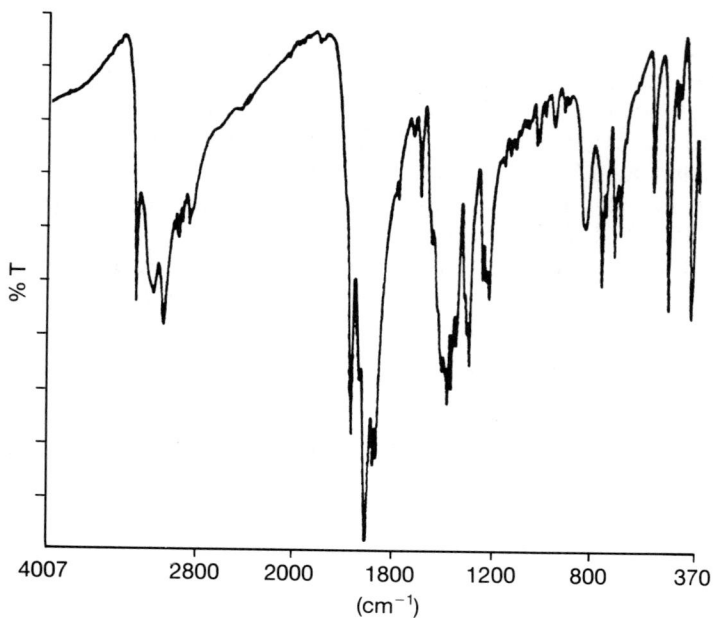

Fig. 8.10 Infra red spectrum of phenobarbitone

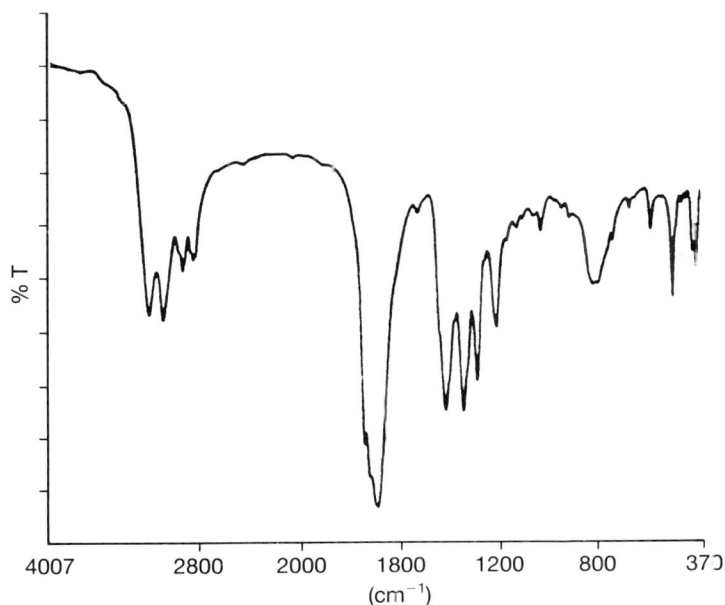

Fig. 8.11 Infra red spectrum of cyclobarbitone

For this reason, the general form of the whole IR spectrum should be considered.

8.5.5 Amphetamines

The infra red spectrum of amphetamine sulphate is shown below.

The broad peak centred at $2938 cm^{-1}$ is due to the N-H stretches. There are a number of bands due to the possibility of the Z N-H bands stretching in, and out, of phase.   Additionally the aliphatic C-H stretches will be responsible for absorption in this region.   The aryl C=C stretches are observed at $1497 cm^{-1}$, 1557^{-1} and 1618^{-1}.

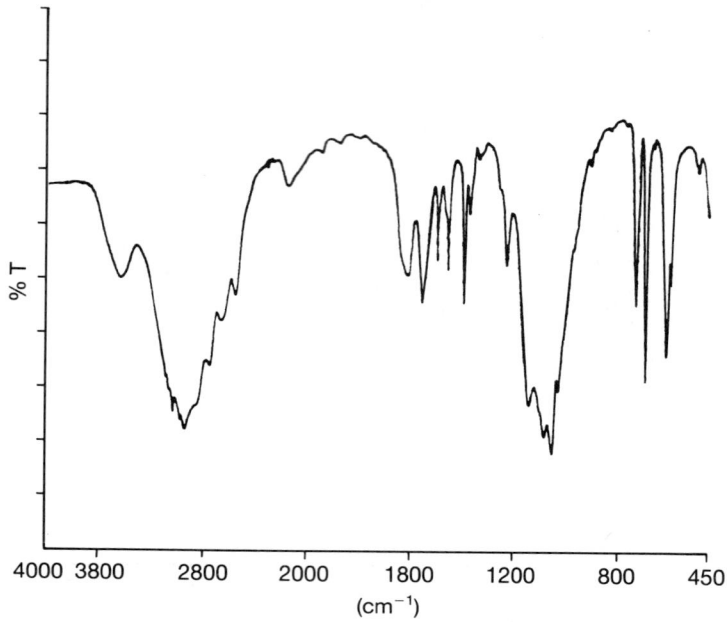

Fig. 8.12 Infra red spectrum of amphetamine sulphate

8.5.6 Benzodiazepines, exemplified by Medazepam

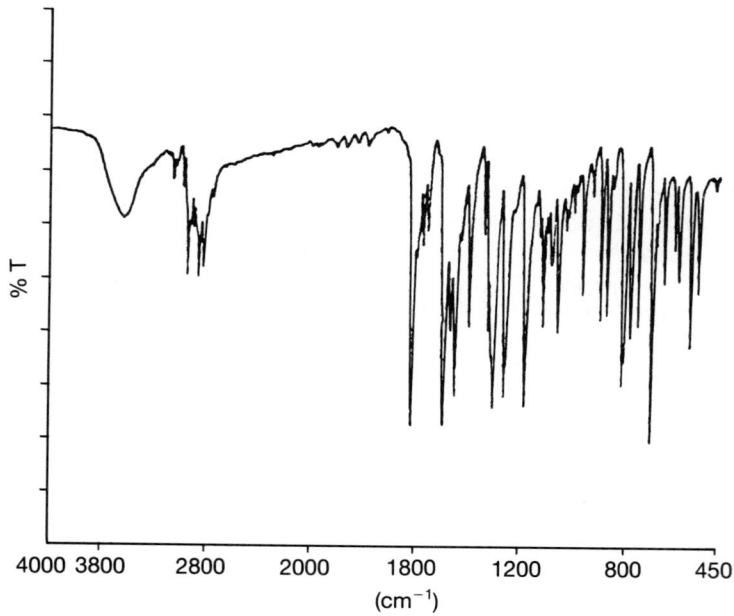

Fig. 8.13 Infra red spectrum of medazepam

The IR spectrum exhibits absorption characteristics of C-H stretches at 2859, 2942 and 2981cm^{-1}. Absorption characteristic of aryl C=C stretches are observed at 1290, 1573 and 1613cm^{-1}.

8.6 Student exercises

Q.1 Develop your own database of infra red spectra, using examples of opiates, cocaine, benzodiazepines, barbiturates and amphetamines. List the principle peaks.

Q.2 You will be provided with some drug samples. They may be mixtures, which require separation. Your demonstrator will explain how this may be done.

Separate any drugs, and identify them using appropriate presumptive tests and infra red spectroscopy.

8.7 Fourier transform infra red spectroscopy

In addition to dispersive infra red spectrometers, a second type of spectrometer is available - the fourier transform infra red spectrometer. These machines have a number of advantages over conventional dispersive instruments. These include :-

1. Quick scanning speed (up to 10 scans per second).

2. They can be used as detectors for gas or liquid chromatographs.

3. A better signal to noise ratio is possible, improving spectrum quality.

4. Microscopy can be coupled to this technique, allowing analysis of very small samples.

5. Mathematical manipulation of the data is possible.

For normal operation, the sample is prepared and mounted in the same way as for a dispersive infra red spectrometer. However, there is a basic difference in the way that the data is collected.

An interferometer (Fig. 8.14) is incorporated into an FT-IR. The basic components of this include a beam splitter, a fixed mirror and a moveable mirror. Half of the infra red light is reflected from the beam splitter to the fixed mirror, and back to the detector, whilst the other half of the light is passed through to the moveable mirror, and then back to the beam splitter, where the beam is again split, and half of the light from the reflected beam is passed to the detector. Using a polychromatic light source, the moveable mirror is moved so that the two light beams interact constructively.

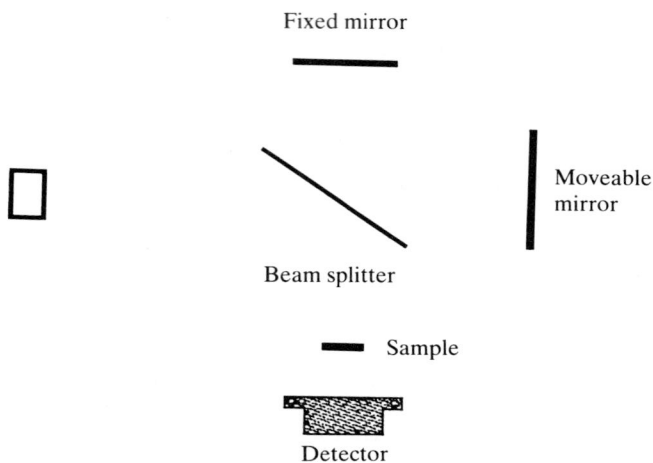

Fixed mirror

Moveable mirror

Beam splitter

Sample

Detector

Fig. 8.14 Schematic representation of an interferometer

The detector records the sum of the sine waves of the polychromatic light, and hence data for all wavelengths to be considered is collected at once, considerably improving the speed of the instrument over the conventional dispersion spectrometer. A sample introduced into the light beam will absorb some wavelengths but not others. Hence some wavelengths will be detected more strongly than others, and an interferogram is produced. These signals can then be turned into an infra red spectrum using a mathematical function called a Fourier Transform. Spectra can also be added together using this function, and hence the signal to noise ratio can be improved.

Further, background subtractions are possible, to eliminate absorbances due to solvents, or matrices in which the spectra are obtained. Additionally, library searches can be obtained.

8.8 Library searches

The compound for which the IR spectrum has been obtained may be identified using a library search. When performing a computerised library search on an infra red spectrometer, a number of factors that will affect data interpretation should be taken into account. Firstly, the spectrum to be considered may have been made under different conditions from those under which the library spectrum was made. For example, a potassium chloride disc may have been used to obtain the library spectrum, but the sample may have been run as a thin film. Thus there may be small shifts in the wavenumbers at which absorbances occur. Where possible, the spectra of the compound in question should be obtained under the same conditions as the library spectrum. It may also be possible to build a library of spectra of drugs likely to be encountered by the analyst.

Secondly, the range of the wavenumbers over which the search is to be carried out should be considered. For example, if a sample is highly hydrated, then very broad -O-H stretching may be observed. However, the sample from which the spectrum in the library was obtained may not have been so highly hydrated, and hence large differences in the spectrum at the high wavenumber end of the spectrum may be observed. It may therefore be preferable to search the library at the low wavenumber end (fingerprint) of the spectrum.

This introduces an additional problem. Whilst the fingerprint is unique to each sample, the intensity of the absorptions at individual wavenumbers may vary. For this reason, and that of slight variation in the absolute wavenumbers at which the absorption occurs, databases and libraries only provide a set of possibilities for the identity of the compound, usually accompanied by a measure of degree of fit of the spectrum obtained with that in the library, covering the wavenumbers over which the search was requested. Thus, whilst FT-IR can be used to gain some idea of the identity of a drug using library searching, the spectrum should always be compared to that obtained for an authentic sample of the compound, obtained under identical conditions.

8.9 Microscopy and fourier transform infra red (FT-IR) spectrometry

Another application of FT-IR lends itself to the identification of drugs. This is through the analysis of drugs using microcrystalline tests, and then confirming the identity of the drug by obtaining the FT-IR spectrum of the crystals formed from the crystal test. This requires a microscope attachment for the FT-IR instrument. This method has been successfully applied to the identification of cocaine [2,3], including samples where the drug has been adulterated with 60% starch and 20% lidocaine.

Using the crystal tests, and the microscope, the crystal type, shape, size and dimensions can be obtained. The crystals are then dried, and using the microscope attachment on the FT-IR, the IR spectrum of the crystals is obtained.

Microscopy has also been applied to the identification of minute quantities of drugs in their crystalline form. For example, lysergic acid diethylamide (LSD) crystals can be identified, after they have been grown on glass wool fibre from a small sample [4].

Thus microscopy, coupled to FT-IR provides a powerful tool in the identification of drugs of abuse.

8.10 References

[1]   Williams, D. H. and Fleming, I.   Spectroscopic methods in organic chemistry,   McGraw Hill, London, 1989, p.264.

[2]   Wielbo, D., Tebbett, I.R., Fitzsimmon, C. and Palenik, S.   "The use of microcrystal tests with fourier Transform Infra red spectrometry for the rapid identification of illicit cocaine", Microgram 23, 1990, pp.258 -264.

[3]   Wielbo, D. and Tebbett, I. R.   "The Use of Microcrystal tests in Conjunction with fourier transform Infra red spectroscopy for the rapid identification of street drugs",   Journal of Forensic Sciences,   37, (4), 1991, pp.1134-1148.

[4]   Harris, H.A. and Kane, T.   "A method for identification of lysergic acid diethylamide (LSD) using a microscope sampling device with fourier transform Infra red spectroscopy",   Journal of Forensic Sciences 36, 1991, pp.1186 - 1191.

9

Gas chromatography - mass spectroscopy of drugs of abuse

9.1 Introduction

Whilst high performance liquid chromatography and gas chromatography provide powerful means of separating the components of complex mixtures, neither technique can identify any component beyond doubt. For example, more than one component may have the same retention time in any given system. This situation is remedied in part by use of two or more chromatographic systems depending on two different physico-chemical processes (i.e. one based on adsorption, and one on partition), although in practice both processes are thought to occur in most types of chromatography.

What is required is another method depending upon another physico-chemical parameter. Diode array detection for HPLC allows some differentiation and purity examination for each eluted peak, but again this technique requires careful use - similar compounds may have chromophores which are difficult to differentiate. More specific detectors are available for gas chromatographs, but again they do not prove beyond question the identity of any given eluate.

Another technique which is available, and in part overcomes these problems, is gas chromatography - mass spectrometry. The gas chromatograph portion of the instrument behaves, and is operated, just like any other chromatograph. However, the detector is replaced by a mass spectrometer (MS), and it is in the sample transfer from GC to MS and the use of the MS as a means of identification of the compound that provide the power of this technique. In simple terms, sample identification relies on the fact that no two compounds would have indistinguishable mass spectra, and that, under identical conditions, compounds fragment reproducibly and predictably and even closely related compounds can be differentiated (Fig. 9.1).

$$[M - CH_2=C=O]^+$$

3-monoacetylmorphine   m/z = 285

$$[M - CH_3CO_2]^+$$

6-monoacetylmorphine   m/z = 268

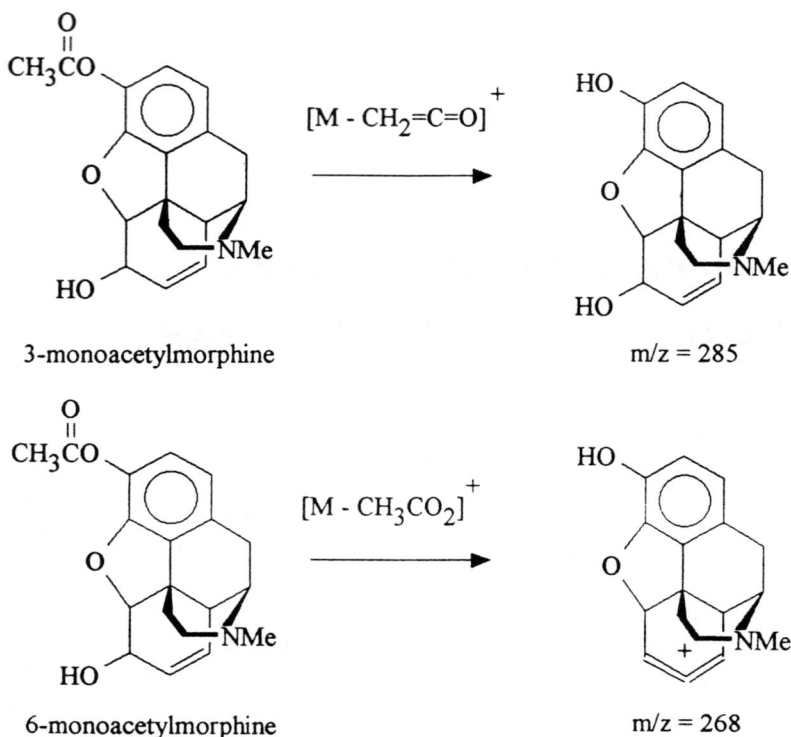

Fig. 9.1 Predicted first fragmentation patterns of
6-mono- and 3-mono acetyl morphines

9.2 The gas chromatograph - mass spectrometer (GC-MS)

The GC-MS consists of a gas chromatograph interfaced with a mass spectrometer.   The gas chromatograph part of the instrument is analogous to a conventional   GC, having the same heated injection block, split/splitless valves, and the same column type present, in a temperature programmable oven.   At the outlet, the column is interfaced with the mass spectrometer. A number of interface types are available   the details of which need not concern us here.   However, they all have one important feature in common, in that they allow the carrier gas to be vented away and the sample to be concentrated prior to entry into the mass spectrometer.

The GC system is modified slightly in that helium rather than nitrogen is used as the carrier gas.   This is because helium is chemically inert, does not interfere with the mass spectral pattern, permits enrichment of the eluent and does not contribute to the ion signal.

The use of the MS requires that a high vacuum be operated for spectral analysis, and so a high powered pump is required.   The system must regularly be checked for leaks (by monitoring for the presence of carbon dioxide, oxygen, nitrogen and water). The mass spectrometer should not be operated if any of these are present in amounts greater than those expected under normal operating conditions. This is because they will damage both the detector and the filament which generates the electrons for the fragmentation process.

9.2.1 Ionization techniques used in drugs analysis

9.2.1.1   Electron impact mass spectrometry

All stable molecules have electrons which occur in pairs. If a molecule is introduced into a vacuum in the vapour phase and bombarded with a stream of high energy electrons, it is possible that one of the electrons from a pair will be knocked out, thereby producing an ion. This ion will be highly unstable, and if the vibrational energy that the ion contains is sufficiently large, certain bonds will vibrate to an extent greater than their elastic limits and the bonds will break, resulting in the formation of daughter ions. These in turn may be in a high energy state and continue to vibrate, which will result in the daughter ions fragmenting still further. It is the ions and the daughter ions that are detected which gives the mass spectrum.

Because of the way that the ions are formed, this technique is called electron impact mass spectrometry. The number and type of ions that are detected depends upon a number of factors, including:

1. the bond strengths of the parent molecule

2. the stability of the daughter ions

3. the energy of the electron beam.

The electron beam is usually set to have an energy level of 70 eV (1 eV = 9.65×10^4 J mol^{-1}). This level is chosen because it will confer sufficient energy on the molecules being analysed to cause them to fragment. If a lower value is used, then the molecule would perhaps only form the molecular ion, whilst if a greater energy value were employed, the ions would all completely fragment.

The ion which forms with the loss of one electron is called the molecular ion. The most intense peak in the mass spectrum is called the base peak. A mass spectrum is really a plot of the frequency of the ions against the mass to charge ratio of the ion.

Electron impact mass spectroscopy has a number of disadvantages. It is difficult to measure the relative molecular mass of some molecules because the parent ion is so unstable. It may not be possible to distinguish between all isomers. Some compounds undergo thermal decomposition so that true mass spectroscopy is not observed.   For this reason, other methods of ionization are sometimes employed.

9.2.1.2   Chemical ionization mass spectroscopy

It is also possible to form ions using charged molecules to strike the molecules being analysed.   Consider the process of chemical ionization (CI) using methane.   Under the correct experimental conditions, it is possible to ionize methane in a stream of electrons:

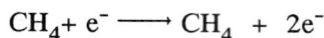

$$CH_4 + e^- \longrightarrow CH_4 + 2e^-$$

The charged methane molecule may then collide with another molecule of methane gas according to the equation:

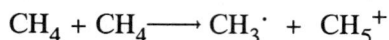

$$CH_4 + CH_4 \longrightarrow CH_3^{\cdot} + CH_5^+$$

The molecule being analysed, (called M for the sake of this discussion) may then collide with the CH_5 formed in the last reaction,   resulting in protonation of the molecule M:

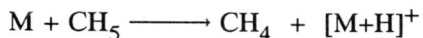

$$M + CH_5 \longrightarrow CH_4 + [M+H]^+$$

Since the protonated ion, written as $[M+H]^+$, has less internal energy than if M had been ionised by a beam of electrons, it is less likely to decompose. This results in a strong signal for the protonated molecular ion.   Controlled conditions must be used to ensure that the electrons used to ionize the methane do not encounter the analyte, and for this reason, there are about 1000 times more methane molecules present than analyte molecules.   Other gases which can be used to ionize the analyte include ammonia and iso-butane.

9.2.1.3   Fast atom bombardment

Using this technique, fast moving atoms of noble gases (for example, xenon) are made to impinge onto the sample, contained in a matrix, on a copper plate.   The ionization of the analyte is brought about by first ionizing the noble gas in a stream of high energy electrons as it passes through an electric field.   The charged noble gas atoms enter a second chamber where they interact with other xenon atoms, to produce fast moving, uncharged xenon atoms.   These atoms then impinge on the sample on a specially prepared copper plate, imparting some of their kinetic energy onto the sample and causing the molecules under analysis to fragment.

Unfortunately, with this system the sample must be dissolved in a matrix (for example, thio-diethanol), which also can lead to the formation of artifacts. In addition, the matrix will have its own mass spectrum. Thus whilst this technique does not lead to the decomposition of some molecules for which E.I. is not suitable, it does have the added complication of the matrix spectrum, and the formation of artifacts. Even subtraction of the matrix mass spectrum will not completely resolve this problem. It should be pointed out that this technique is not widely used in drug analysis.

9.2.1.4 Advantages and disadvantages of mass spectroscopy

Whilst this technique means that a compound in a mixture may be identified through use of gas chromatographic retention time and mass spectrum, either in comparison to a library of such data, or through comparison with a standard sample, there are a number of difficulties associated with this technique. The complex matrix in which the compound of interest is found may interfere with the chromatogram and mass spectrum. This can be overcome by subtracting the background noise if this facility is available, but this means that some data may also be lost.

Compounds can be identified through comparison of the mass spectrum obtained with that from a library, or with control compounds analysed on a previous occasion. This, however, makes the assumption that the library data and experimental data were obtained under identical conditions, which is frequently not the case. This too would determine the relative frequencies of many of the ions encountered in the spectrum. These considerations will be illustrated throughout this course, but in practical terms mean that a sample must always be compared to a control compound under identical analytical conditions.

Further advantages of this technique include the availability of a variety of ionization techniques, which reveal different kinds of information, although only electron impact techniques (the most widely used) will be considered in this study. Further advantages include the possibility of selected ion monitoring (obtaining chromatograms of ions characteristic of the compound of interest) and the ability to quantify a sample using deuterated analogues of compounds as internal standards.

Disadvantages include the cost of the equipment and its maintenance, and the requirement for technical know-how.

9.3 Data interpretation

A number of pieces of information may be obtained from a GC-MS.

9.3.1 Total ion chromatograms

This is a plot of the total number of ions (Y axis) plotted through time (X axis), and provides a trace which is considered analogous to the trace from a normal gas chromatograph. An example of this is shown below (Fig. 9.2) for the separation of the TMS derivatives of five opiate drugs.

This is analogous to the chromatogram produced by a flame ionisation detector. It will provide information about the retention time of a compound, but nothing about the structural details of the components of the mixture.

9.3.2 Mass spectral data

9.3.2.1 Mass spectra

In addition to the retention time of a compound, obtained from the total ion chromatogram, it is possible to obtain a mass spectrum for each of the scans. This will also give a mass spectrum for each of the eluted peaks. This is illustrated (Fig. 9.3) for the diamorphine, the last peak obtained in the chromatogram shown above. The mass spectrum is a second means of identifying the eluted compound.

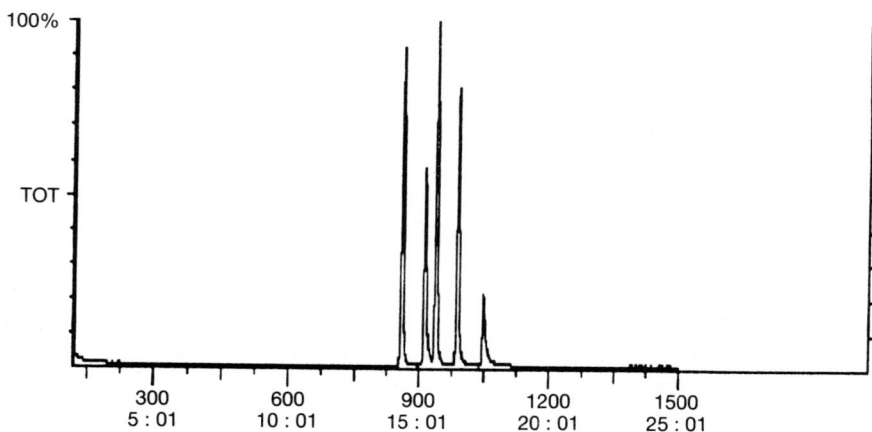

Fig. 9.2 Total ion chromatogram of five opiate drugs chromatographed as
TMS derivatives in the system described (section 9.4.1.1).
Elution order: Codeine, acetylcodeine, morphine,
6-0-monoacetylmorphine, diamorphine

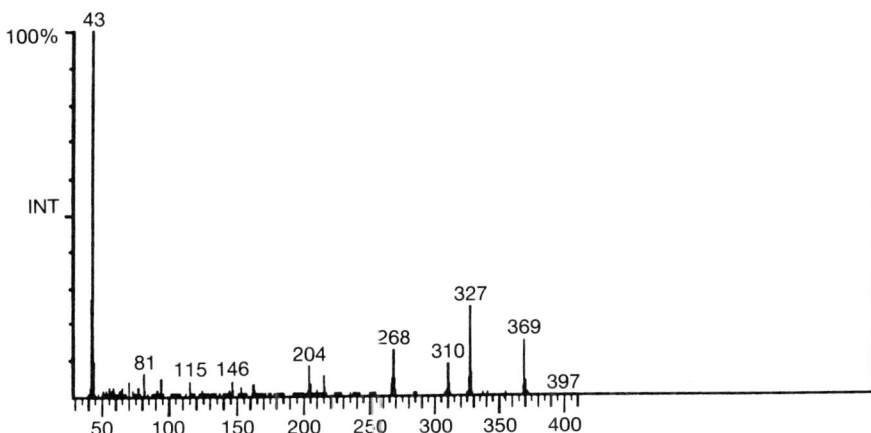

Fig. 9.3 Mass spectrum of diamorphine separated from other
opiates in the system described (section 9.4.1.1)

Sample identification can be achieved through the comparison of the mass spectrum with a library mass spectrum. Such a library is available from the National Bureau of Standards.   However, there are problems with this method of sample identification.   The quality of the mass spectrum obtained depends upon a number of factors.   Of particular importance is the degree of resolution of the compounds as they elute from the gas chromatograph.   The resolution of an opiate sample, without derivatisation, is shown in Figure 9.4. It is clear that the peaks overlap and are very broad.   Resolution has not been achieved.   This means that the use of mass spectroscopy is almost pointless since the mass spectrum will be that of a mixture.

The amount of material which "bleeds" from the column into the mass spectrometer also influences the quality of the mass spectral data.   In the total ion chromatogram this phenomenon is observed as a drifting baseline. Column bleed may be due to decomposition products retained on the chromatographic column which will elute as the temperature of the column increases.   Column bleed increased the amount of noise in the mass spectrum, and becomes especially problematic if the drug of interest is only present at very low concentrations.   A noisy mass spectrum can also cause difficulties with sample identification.   An additional contribution to the problem arises from poor column stability which causes the stationary phase to bleed into the mass spectrometer at high temperatures.

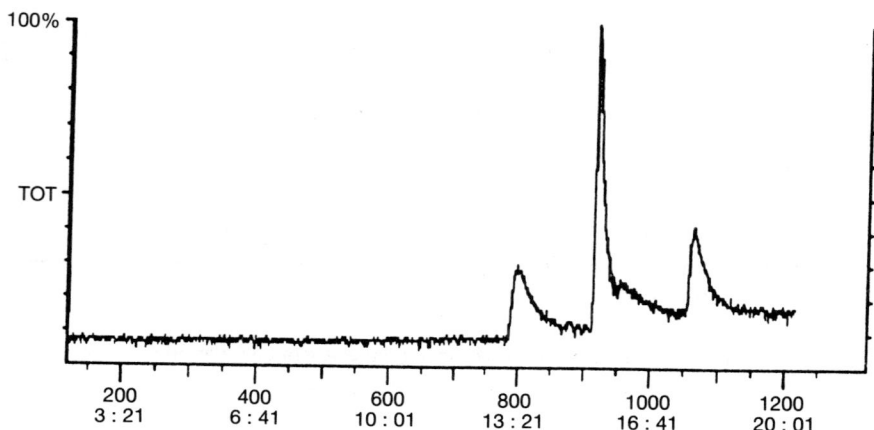

Fig. 9.4 Example of unresolved opiate drugs for which
mass spectrometry would not reveal useful information

If the mass spectrometer detector becomes dirty, the sensitivity of the instrument is reduced. One way to prolong the life and cleanliness of the detector system is to avoid use of the filament until after the solvent front has eluted from the system. This prevents a large amount of contaminated material becoming ionised and contaminating the electron multiplier.

Due to these problems it is common practice to run a standard sample of the suspected drug through the GC-MS system prior to analysing the sample. This will permit a retention time and mass spectrum to be obtained under as near identical conditions as possible for both sample and standard. It is also important to consider the relative peak heights of the fragments observed, since a compound should fragment reproducibly under a given set of experimental conditions. If the retention time, mass spectrum and relative intensity of the observed ions are correct, for a given compound, then it is almost certain that a positive identification can be made.

The mass spectrum can also be interpreted in terms of the fragmentation pathway of the molecule. This is shown below (Fig. 9.5) for the fragmentation pathway of diamorphine.

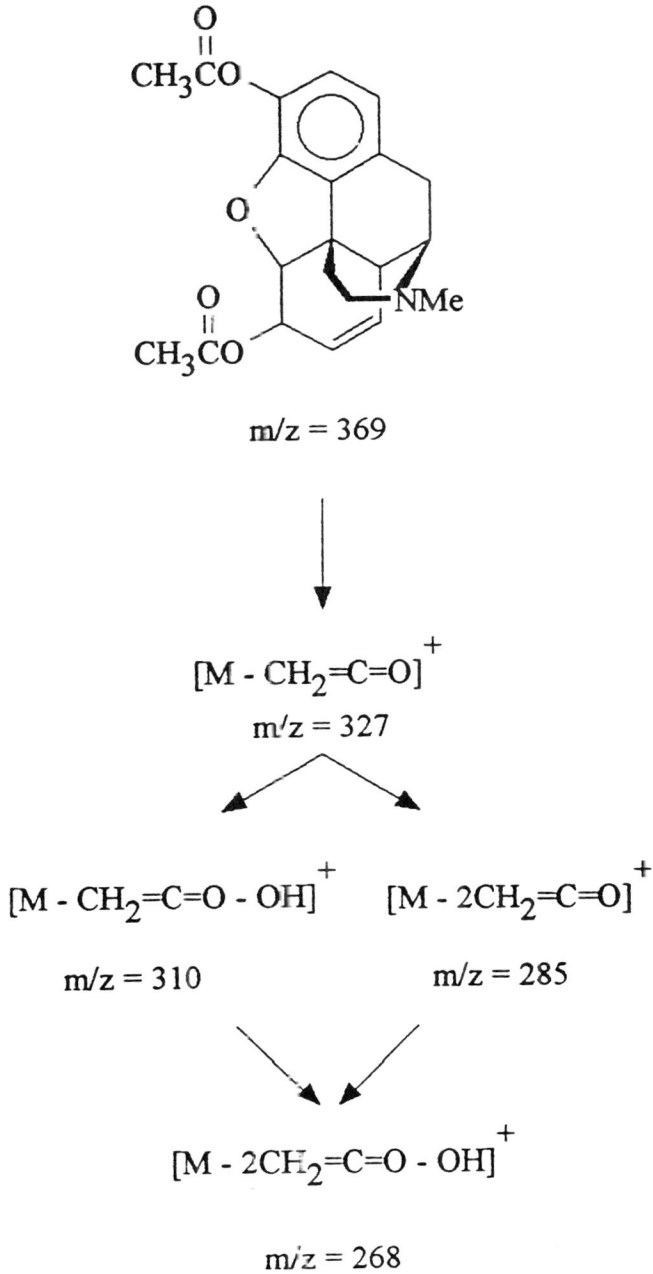

Fig. 9.5 A possible mass fragmentation pathway of diamorphine

9.3.2.2 Isotope peaks

The mass spectrum will also contain fragments with values of 1 or 2 mass units above or below the major fragments.   Consider for example the fragments around the m/z = 369 ion of diamorphine, shown below.

Those at one mass unit higher than the molecular ion occur due to the presence of some molecules which contain either a ^{13}C atom or ^{2}H atom, or have been protonated in the mass spectrometer.   More rarely, molecules with two of these rare isotopes will occur, and these will have a relative molecular mass two units higher than the molecular ion (m/z = 371).   The ions at m/z 367, 368 are due to deprotonated ions which have lost either one, or two protons respectively.

Since each compound under a given set of conditions will fragment in a certain way, then the relative intensity of the different fragments should also remain constant.   The most intense ions can be compared to tabulated lists of the most prominent ions and this may afford another means of sample identification.   Again, it should be pointed out that the lists are best prepared by the scientist on the machine which they will be using, since the intensity of the ions may vary a little between machines, and under different operating conditions.   Additionally, it should also be remembered that the mass spectra obtained from drugs after derivatisation will be different to the underivatised samples.

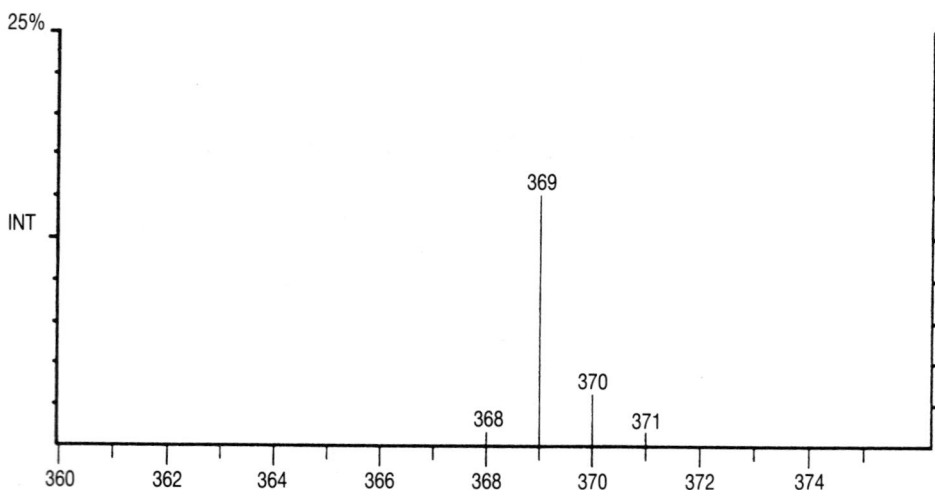

Fig. 9.6 Isotope peaks observed around the molecular ion
of diamorphine (m/z = 369)

9.3.2.3 Selected ion monitoring

If the drug sample being analysed is complex, it may not be possible to obtain baseline resolution for the different components of the mixture.   This is exemplified by the chromatogram below, which shows a separation achieved for cannabis resin, after derivatisation with N,O-bis trimethylsilyl acetamide (Fig. 9.7).

As can be seen, the mixture was complex, and it was not possible to identify Δ^9-tetrahydrocannabinol by examination of the retention times alone. However, if the appearance of two of the prominent ions observed under these conditions for Δ^9-tetrahydrocannabinol after derivatisation are observed (m/z = 386, the molecular ion, and m/z = 375 [M - Me]$^+$), it can be seen that they occur together (Fig. 9.8), at the correct retention time for the parent compound, Δ^9-tetrahydrocannabinol.   This indicates that this compound is indeed present in this sample and that this sample is indeed a cannabis product.

This method is fraught with problems.   The selected ions only form a small part of the ion count. If the sample is only a very weak one, then the selected ions may not be visible amongst the background noise.. Additionally ions must be chosen which are characteristic for the drug in question.   In this example, it can be stated that the compound is Δ^9-tetrahydrocannabinol and not the isomer, Δ^8-tetrahydrocannabinol, because of the retention time.

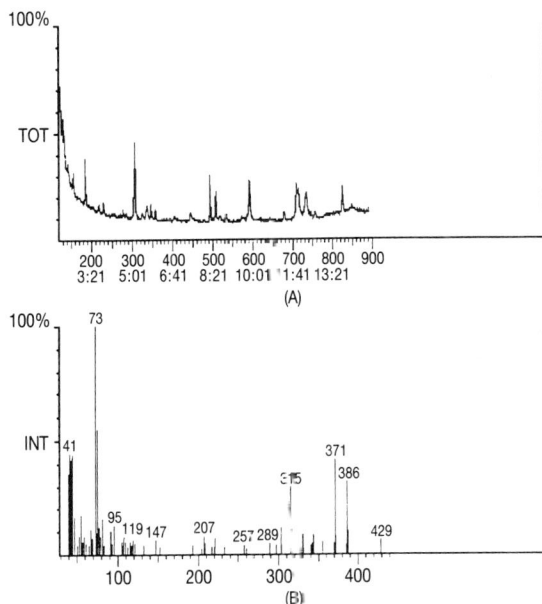

Fig. 9.7 Separation of compounds in a cannabis resin sample after derivatisation with BSA (A) and mass spectrum of standard Δ^9-tetrahydrocannabinol derivatised with BSA (Retention time = 13.75 min) (B)

9.3.3 Quantification using deuterated internal standards

If the mass spectrometer has sufficient resolution, it is possible to use deuterated internal standards to quantify the amount of drug in a mixture. Some deuterated standards are commercially available, but others can be synthesized using deuterated precursors - for example d_6-diamorphine using morphine and deuterated acetic anhydride.

The method is performed as follows:

A series of standards at known concentrations are prepared in a solution of the same **deuterated** compound, the internal standard being at the same concentrations for all of the samples to be analysed. These are then chromatographed on the GC-MS. The total ion count is obtained for suitable ions for quantification. For example, suitable ions from the mass spectrum of diamorphine are m/z = 369 and it's deuterated analogue m/z = 375 (molecular ions) and m/z = 327 and m/z - 331 (first fragmentation products) (Figs. 9.9 and 9.10). At each of the concentrations of the standards, the following ratios are calculated:

$$\frac{\text{Peak area m/z} = 369}{\text{Peak area m/z - 375}}$$

and

$$\frac{\text{Peak area m/z} = 327}{\text{Peak area m/z} = 331}$$

The **ratios** are plotted against the concentrations of the standards, and using regression analyses the equation of the straight line obtained. The sample of unknown concentration is dissolved at a suitable concentration in the solution of deuterated internal standard at the same concentration as that used to prepare the calibration curve and is chromatographed in the same system. The ratios of the peak areas for the chosen ions and the deuterated equivalents are obtained, and from the rearranged regression equation, the concentration of the drug in the sample is calculated.

When performing this method, all the usual precautions in preparing a calibration curve should be observed. These include running sufficient samples, ensuring that the lowest concentration of standard is chromatographed first, and that blank injections are made between each of the samples to ensure that there is no carry over. An example of the type of data which can be expected is illustrated below.

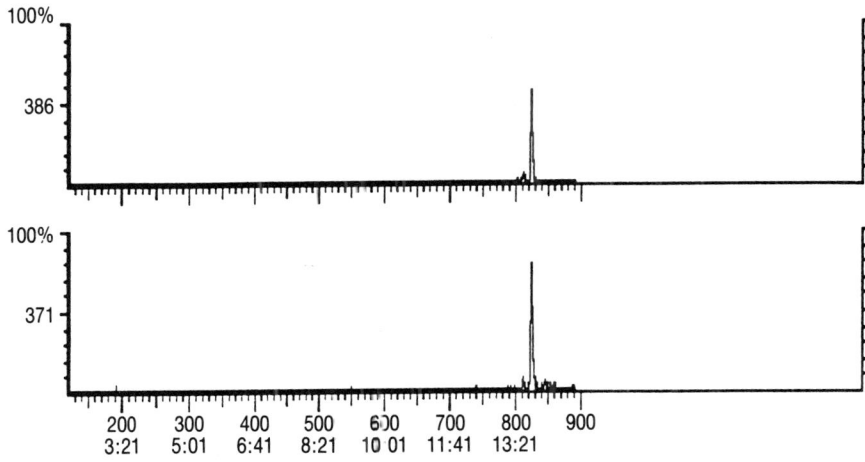

(A)

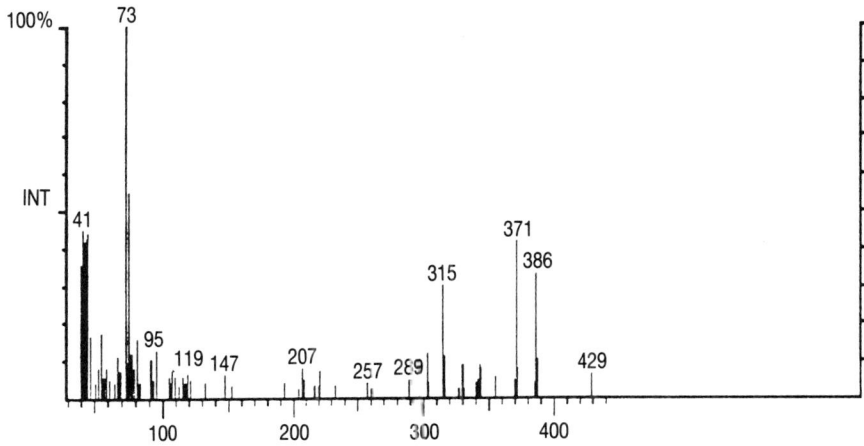

(B)

Fig. 9.8 Selected ion chromatogram of a cannabis sample,
monitoring m/z = 386 and m/z - 371 (A) and mass spectrum at that
retention time (13.74min) (B)

Using a range of sample concentrations, calibration curves can be
obtained as exemplified below (Figs. 9.11 and 9.12).

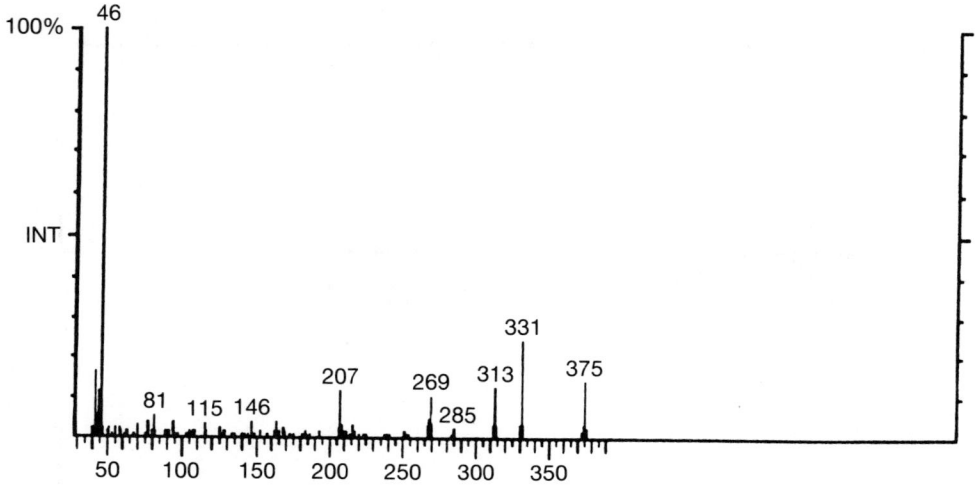

Fig. 9.9   Mass spectrum of hexadeuterated diamorphine, used as an internal standard

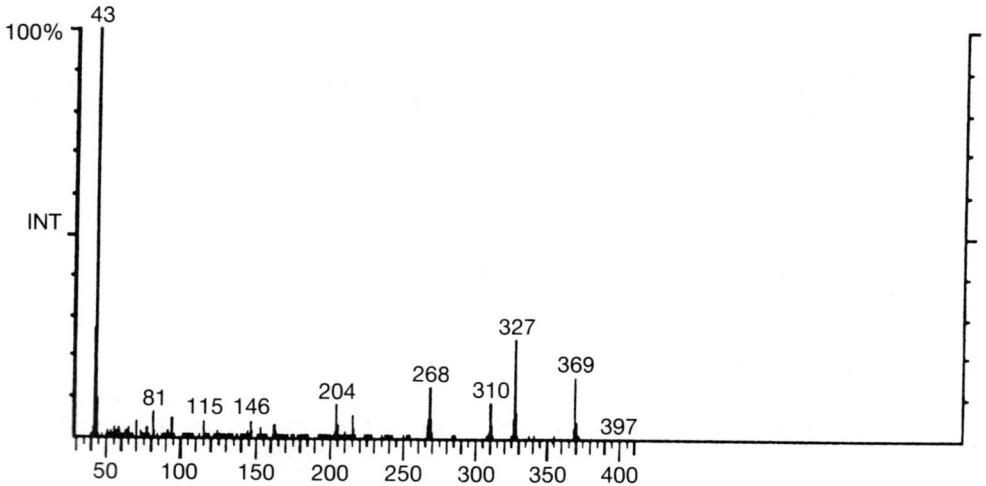

Fig. 9.10   Mass spectrum of diamorphine showing ions which may be used in conjunction with deuterated internal standards for quantification

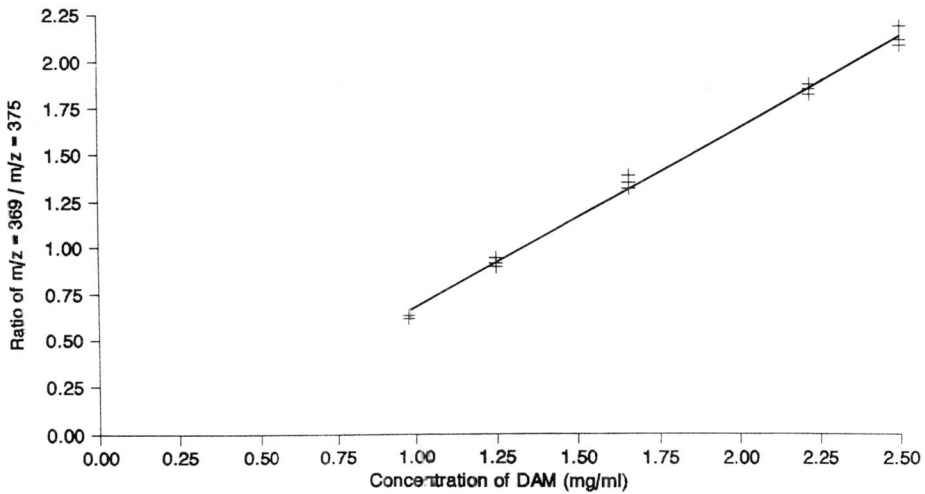

Ratio = 0.98 x Conc (mg/ml) - 0.3

Fig. 9.11　Ratio of peak area of deuterated diamorphine to
diamorphine using m/z = 369 / m/z = 375

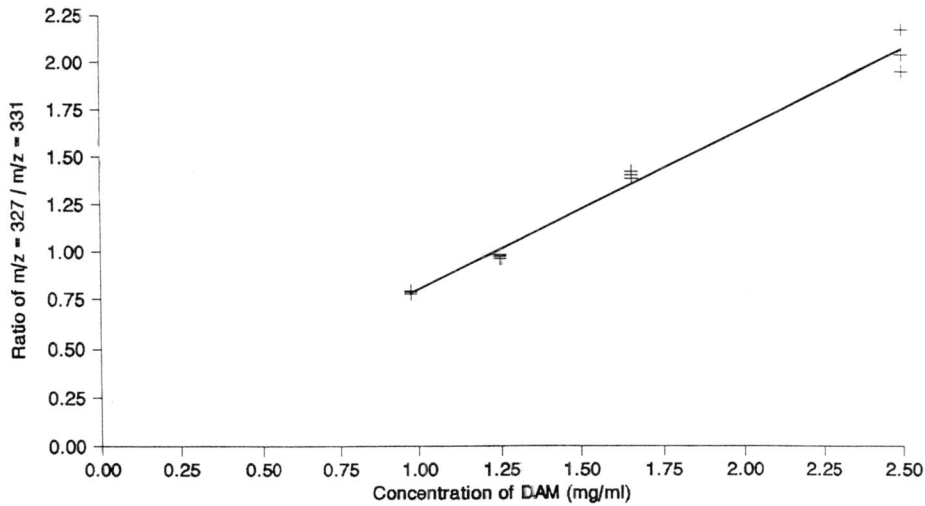

ratio = 0.86 * conc. (mg/ml) - 0.06

Fig. 9.12　Ratio of peak area of deuterated diamorphine to
diamorphine using m/z = 327 / m/z = 331

From the regression equations, the concentration of diamorphine in an illicit sample can then be calculated, in an analogous way to that using an internal standard and flame ionization detector.

9.4 Student exercises

9.4.1 Underivatised drugs

9.4.1.1 Opiate drugs

Using the programme described below, chromatograph each of the following; codeine, morphine, acetylcodeine, 6-0-monoacetylmorphine, and diacetylmorphine.

Column:	25m BP-5 0.25μm thickness 0.2 mm i.d.
Temperature:	200°C for 2 min, the to 280°C @ 6°C/min, hold for 3 min.
Injection Vol:	1 - 3μl
Carrier Gas:	Helium, 1 ml/min
Split ratio:	60 : 1

Q.1 Examine each of the chromatograms obtained. Determine the retention time of the compounds. Once you are satisfied with the chromatography, obtain the mass spectrum of each of the compounds. Determine what ions are present and how they they have arisen.

Q.2 Attempt to identify your kown compounds from the NBS library, stored on the computer of the mass spectrometer (you will be shown how to do this). What problems do you encounter, and how may they be overcome?

9.4.1.2   Cocaine and metabolites

Using the GC system described below, chromatograph cocaine on the GC-MS.

Column:	OV - 1 or BP - 5, as above
Temperature:	170°C for 2 min, then to 280°C @ 16°C/min hold for 10 min.
Injection Volume:	1 - 3µl
Carrier Gas:	Helium 1ml/min
Split Ratio:	60 : 1

> **Q.3** How does the mass spectral data that you obtain compare with the spectral library?
>
> **Q.4** Does your data match the predicted breakdown pathway of cocaine, shown in Figure 9.13?

9.4.2 Derivatised samples

In addition to the analysis of samples of drugs of abuse in their underivatised state, it is also possible to chromatograph the derivatives of such samples. The advantages of this are that the chromatography and resolution of components of mixtures are often improved, whilst the compounds will still have characteristic fragmentation pathways. These are, however, different to the underivatised samples. However, unless the derivative of the sample compound can be compared to the derivative of an authentic marker, or one which has been stored in the library, then the library searching facility is lost.

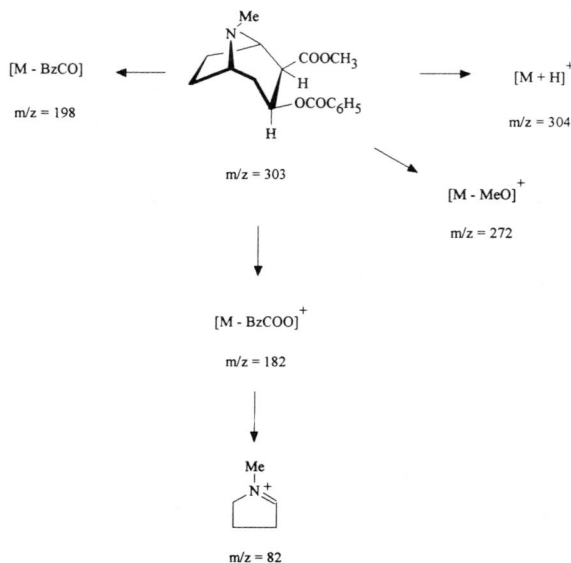

Fig. 9.13 Fragmentation pathway for cocaine under
electron impact mass spectroscopy

9.4.2.1 Opiate drugs

> **Q.5** Derivatise each of the opiates and run on the GC-MS as before.
> Using the chromatograms and MS data, interpret these in
> terms of possible fragmentation pathways. Compare the
> resolution with that obtained from underivatised compounds.
> Using samples prepared to simulate street seizures that you are
> likely to encounter in your home country, examine the
> chromatography of the mixtures both as underivatised
> compounds and after derivatisation. Which is better, and why?

9.4.2.2 Cannabinoid drugs

Q.6 Devise a temperature programme for the analysis of cannabinoid drugs. Obtain the retention time and MS data for the TMS derivatives of THC, CBN and CBD.

Q.7 You are also provided with leaf and resin samples of material from real <u>Cannabis</u> seizures. Identify the compounds present, using the MS data of the TMS derivatives.

9.4.2.3 Drugs of particular interest

Q.8 You will be able to discuss with your tutor the particular drugs with which you are working in your home country. Dertermine whether GC/MS will aid in their identification and quantification. Attempt to devise GC/MS methods that you will be able to employ in your own laboratory.

9.4.2.4 Quantification using deuterated internal standards

Q.9 Using the method described above for the quantification of diamorphine in illicit samples, identify and quantify the amount of diamorphine and its surrogates in the powders with which you are provided.

10

Student exercises: answers

Chapter 2 Physical descriptions of drugs of abuse

A2.1 Fresh plant material of <u>Cannabis sativa</u> has a strong odour, slightly reminiscent of mint. The smell becomes stale after the plant material dries out. There is a difference in the trichome density on different parts of the plant. The trichome density is greatest on the ribs of the leaves, and around the female flower parts. There are also trichomes on the stems, and the surfaces of the leaves. The density on the leaves is greater on the underside than on the upper surface of the leaves. The glandular trichomes occur with the greatest density on the female flower parts, although they do occur on other flower parts. The non-glandular trichomes all point upward on the stems.

A2.2 There are some glandular secretions at the base of the sepals of the hop flowers. If they were examined by the inexperienced eye, it is possible that they could be confused with the glandular trichomes of <u>Cannabis sativa</u>.
     Tobacco has some trichomes present, although no glandular trichomes are observed. Additionally, tobacco has a completely different smell to Cannabis, and a different colour.

A2.3 You will see plenty of trichome material, and the broken parts of glandular trichomes. If you are lucky, you will still manage to see some of these intact.

A2.4 The principle problem of examining this type of material is that it will be discoloured, and may have shrunk or dried out since seizure.

A2.5 The first thing which should be examined and described is the wrapping material that may be found on the block. Details which should be included are colour and weight of the block. General texture of the block should be noted visually and under the microscope. A dark crust may be present around a lighter interior. The thickness of this "crust" should be recorded.

The surface may be marked with striations or indentations from the packaging. These should be recorded, and if possible, photographed. These striations may be used to illustrate that different pieces of resin may once have originally been a single block.

A2.6 The colour and texture of the crystals making up the powder should be recorded and their dimensions measured. Any oil present may have come from the mother liquor from which the crystals were isolated. The colour of this oil should also be noted. Similarities and differences in such details may be used together with other evidence in determining whether drug seizures have a common source.

Although this list of observations is by no means exhaustive in giving you the opportunity to record physical characteristics, you should now be in a position to write, and understand the importance of a good physical description. This is non-destructive, an important consideration where the sample size is limited, but at the same time may reveal important information concerning the sample.

Chapter 3   Presumptive tests for drugs of abuse

A3.1 You will obtain weak positive results from the herbs and spices provided that they are really fresh. This is because this reagent reacts with a large number of plant phenolic compounds, and is not specific to cannabinoids. However, the colour reaction may well be slower and less well defined than that obtained from <u>Cannabis</u> products. Confusion could arise if the <u>Cannabis</u> products were old, or the cannabinoid content low.

A3.2 The colour reactions that you should obtain are listed in Table 3.2.

A3.3 Some slight variation may be present because of (i) the presence of impurities in the drugs, (ii) the perception of colour varies between individuals, and so it is always necessary to perform both positive and negative controls.   Additionally, colours may be masked as in highly pigmented samples containing plant phenolics.

A3.4 These results indicate that it is possible to say that a drug giving a positive result may be an opiate, but it is not possible to identify which specific compound is present.

A3.5 The variety of colours that you should obtain are detailed in Table 3.3.

A3.6 Cocaine and benzoylecgonine react with the sodium hydroxide in methanol to yield methyl benzoate also know an "oil of wintergreen" which has a characteristic smell.   The production of both colours from the different reagents plus the characteristic smell of "oil of wintergreen" are strongly indicative of the presence of cocaine or its breakdown products.

A3.7 The pure drugs give strong reactions producing a purple colour. Tablets should give a similar response unless any reaction is masked by the presence of dyes.

A3.8 Whilst microscopy and spot tests cannot be definitively used to identify a compound, they are widely used to determine the class of compound to which the drug belongs.   For example, a positive result in the Duquenois Levine test coupled to a thorough microscopic examination is an excellent indicator for the present of <u>Cannabis</u> plant material.   A positive result with cobalt isothiocyanate and the release of methylbenzoate (the smell of "oil of wintergreen") is an excellent indicator for the present for cocaine or benzoyl ecgonine.

Chapter 4 Thin layer chromatography of drugs of abuse

A4.1 The following data have been obtained using the system for cannabinoid analyses. [Table 10.1].   The compounds may be identified on the basis of both R_f value, and colour reaction with the spray reagent.

Table 10.1 Data obtained from TLC analysis of <u>Cannabis</u> products

Drug	R_f Value	Colour reaction with Fast Blue B
Cannabinol	0.51	Mauve
Δ^8-THC	0.69	Orange
Δ^9-THC	0.69	
Cannabidiol	0.79	Light Brown

A4.2 A number of problems are encountered during compound identification. These include co-elution (for example that of Δ^8 and Δ^9-THC. Further, the extremely complex nature of <u>Cannabis</u> products makes positive identification by TLC very difficult.   In older samples of <u>Cannabis</u> the THC concentrations may be extremely low, and these compounds difficult to detect.

A4.3 As the solution prepared from <u>Cannabis</u> products ages, so the amount of THC decreases, and that of the breakdown products increases.   This means that once the extract of the <u>Cannabis</u> product has been prepared, it should be analysed as soon as possible, especially if comparisons between samples are to be made.   Ethanol is the most suitable solvent for preparation of extracts since the cannabinoids appear to be most stable in this solvent.

A4.4 The ammonia is added to the second solvent since these compounds are all bases.   Under the basic conditions provided by the ammonia, the drugs occur mainly in the uncharged form. This reduces the amount of tailing which can be seen under the condition of straight phase chromatography and improves the chromatography of the analytes.

A4.5 The R_f values observed (Table 10.2) can be explained as follows. The least polar compounds, diamorphine and acetylcodeine both have substituents on both hydroxide moieties. These will interact less strongly with the silanol groups of the silica gel - some interaction will still occur due to the polarity of the double bonds and carbonyl groups present. The compounds with one hydroxide substituent have intermediate R_f values, since they will also have an intermediate polarity. The most polar compound, morphine, migrates the least because this compound forms the strongest bonds with the stationary phase.

A4.6 Whilst the R_f values will be close to those given in the table, they will not be identical. This can be explained because the temperature of the analytical system will be different, as will the vapour saturation of the tank. Increased temperature results in higher R_f values being obtained, as does increasing saturation of the TLC tank with solvent vapour. The converse is also true.

Table 10.2 R_f values obtained for opiate drugs under the TLC conditions described

Compound	Solvent System	
	Chloroform/ Methanol 9/1	Ethyl acetate/Methanol/ Ammonia 85/10/5
Morphine	0.04	0.32
Codeine	0.19	0.47
6-monoacetylmorphine	0.21	0.63
Acetylcodeine	0.32	0.74
Diacetylmorphine	0.32	0.74

A4.7 The observations you have made mean that standards should always be run alongside the sample being analysed, so that comparison can be directly made between sample and standard.

A4.8 As the standard solution of diamorphine ages, so 6-0-monoacetylmorphine and then morphine will appear. This is due to the hydrolysis of the acetate functionalities. Diamorphine is stable in the dry, solid state but <u>not</u> as a solution. For this reason standard solutions and samples should be prepared immediately prior to use.

A4.9 The observed Rf values (examples are shown in Table 10.3) can be explained in terms of the substituents found on the carboxylic acid and hydroxide moieties of the ecgonine. Ecgonine has no substituents and will therefore form the strongest interaction with the stationary phase. This will result in the lowest Rf values. Benzoylecgonine has one substituent, but the benzoyl moiety has a number of delocalised pi-electrons in the benzene ring, which are able to form a strong interaction with the stationary phase. The carboxylic acid moiety is also strongly polar. Ecgonine methyl ester has only one substituent, but the methyl ester is not so polar as the benzoyl moiety, and so this compound migrates further. Cocaine has substituents on both hydroxide groups. These interact with the stationary phase (otherwise the compound would run with the solvent front), but less strongly than the hydroxide groups so this compound has the highest R_f values.

Table 10.3 TLC data obtained using the TLC system described

Drug	Rf value
Cocaine	0.71
Ecgonine methyl ester	0.52
Benzoyl ecgonine	0.32
Ecgonine	0.25

A4.10 The principle problem with this technique is lack of sensitivity.   This can be overcome in part by use of TLC plates containing a fluorescent indicator, and observing under UV light at 254nm.   The compounds will appear as dark spots on a pale background, whilst they will fluoresce at 360nm.

Another way around the problem is to spray the TLC plate with sulphuric acid and heat it to 100°C for five minutes.   There is, however, the obvious difficulty of spraying an acid, and extreme caution should be exercised.   Further, this reagent is very non specific.

A4.11 It should be possible to identify the drugs present in the tablets by comparison of the R_f values of the compounds extracted with those of standards.   A problem may arise if the R_f values for the drugs are very close and the compounds are not well resolved.   Further saturation of the TLC tank with solvent vapour should help overcome this problem.

A4.12 The differences in R_f value can be explained in terms of the substituents found at C_5.   In general, increasing the carbon chain length results in a decrease in polarity, and increased R_f value.   However, by cyclyzing the chain, a more polar substituent is created, which then has a lower R_f value.

The difference between phenobarbitone and cyclobarbitone can be explained in terms of the polarity of the six membered carbon ring. The cyclobarbitone contains one double bond, whilst the ring electrons in phenobarbitone are delocalised around the whole resulting in an increase in polarity. This means that phenobarbitone will be more strongly retained than cyclobarbitone, hence the observed differences in Rf values.

Chapter 5 - Gas chromatography

A5.1 The elution order of the opiate drugs is codeine, morphine, acetylcodeine, monoacetylmorphine and diamorphine. This can be explained as follows. The gas chromatographic system separates the compounds on the basis of polarity, with the drugs partitioning between the stationary phase and the mobile phase. On a non polar column, as used here, the most polar compound will elute first, and the least polar compound last. Whilst one would expect morphine to elute first, it forms very strong hydrogen bonds and does not chromatograph well. This is why codeine elutes first, followed by morphine. Acetylcodeine has a 6-acetyl group and so is slightly more lipophilic, slightly more volatile and has a higher molecular weight. Monoacetylmorphine does not have the 3-methyl ether, and so is slightly less volatile, and elutes next, whilst diamorphine, the most lipophilic, has the highest molecular weight and elutes last. It can therefore be seen that the degree of lipophilicity, the effect of the presence, or not, of hydroxide groups on volatility, and the molecular weight, all contribute to the chromatographic behaviour of the analytes.

A5.2 The values will be close to those quoted in the literature. Differences are due to the age and quality of the GC column, and the exact experimental parameters used for the separation process.

A5.3 Poor chromatographic behaviour of underivatised drugs, and the complex nature of some street samples means that sample identification under these GC conditions is extremely difficult, and sometimes a positive identification cannot be made.

A5.4 You will notice that tailing of some compounds, and poor chromatography (e.g. for morphine) still present problems, as does co-elution. This can be overcome by derivatisation of the sample, and using a temperature programme.

A5.5 The improved chromatography of morphine can be explained by examining the structure of morphine, and the derivatisation process

involved.   Morphine has two hydroxide moieties, both of which
contribute to the polarity and strong hydrogen bonding ability of the
molecule.   Derivatisation of the molecule with BSA results in these
highly polar moieties becoming derivatised to trimethylsilyl ethers, which
exhibit much improved chromatographic behaviour (Fig. 10.1).

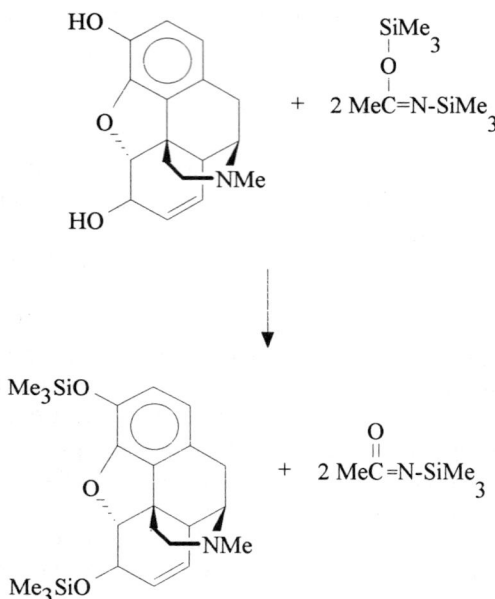

Fig. 10.1 Reactions involved with derivatisation of morphine with
N,0-bistrimethylsilylacetamide

A5.6 It is not possible to definitively identify each component of this
Cannabis resin because baseline resolution cannot be achieved.   One
way to overcome this problem is to use GC-MS and HPLC in the
analysis of the resin sample.

A5.7 The surrogates all possess very highly polar hydroxide and carboxylic
acid moieties, which result in these compounds forming very strong
hydrogen bonds, and poor chromatography.   Again, by derivatising
these functional groups, molecules with better chromatographic
properties are obtained.

As the chain length of the substitution at C_5 increases, so does
retention time.   The difference in the retention time of phenobarbitone
and cyclobarbitone can be explained in terms of the relative polarity of
the substituent at C_5, the more polar phenobarbitone eluting prior to the
cyclobarbitone.

A.5.8 The type of chromatogram obtained is illustrated below (Fig. 10.2.).

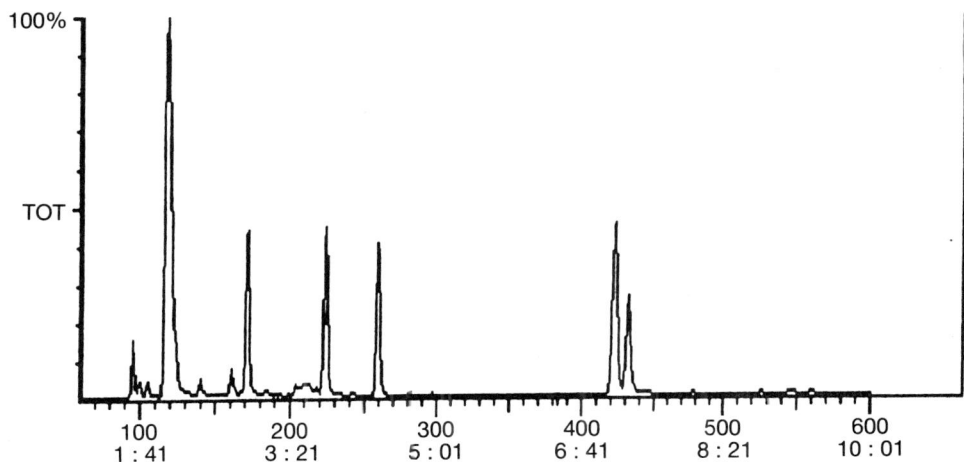

Fig. 10.2 Separation barbiturate drugs after derivatisation with 0.2M
trimethylanilium hydroxide in methanol.   Elution order;
barbitone, butobarbitone, amylobarbitone, pentobarbitone,
phenobarbitone, cyclobarbitone.

Chapter 6   High performance liquid chromatography of drugs of abuse

A.6.1 An example of the separation achieved is shown in Figure 10.3, along with the structure of the molecules in question. The first compound to be eluted is acetylcodeine. This is the least polar. The next, diacetylmorphine has a more polar substituent (acetate versus methyl ether) at the 3 position. 6-O-monoacetylmorphine elutes next. The phenolic group at the 3 position is more polar than the methyl ether or the acetate group of the 2 compounds eluted prior to this. Codeine elutes next. The alcohol at the 6 position is very polar, with the 3-hydroxide moiety replaced by a methyl ether. The last compound to be eluted is morphine, with two hydroxyl groups, the most polar of the compounds analysed. The increasing polarity leads to increasing interaction with the stationary phase and hence longer retention times.

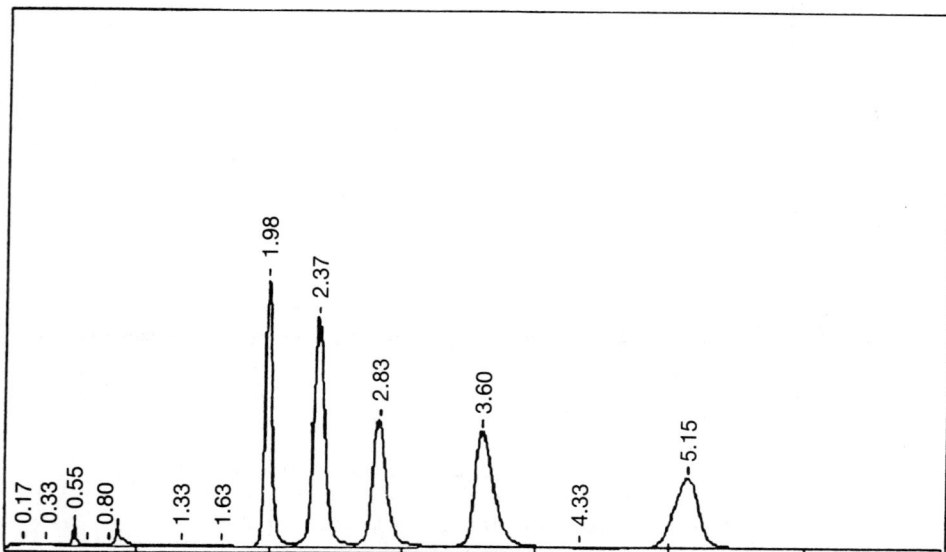

Fig. 10.3 Chromatographic separation of the opiates by HPLC (elution order; acetylcodeine, diamorphine, 6-0-monoacetylmorphine, codeine, morphine)

A.6.2 Baseline resolution is necessary because when quantification is applied to any one compound, the recording device must assign a peak height or peak area to that compound alone. If baseline resolution is not achieved, then the recorded measure will not represent that compound alone, and will not be accurate.

A.6.3 Linearity must always be established and a new regression equation calculated every time a new batch of solvent is used, or another part of the HPLC system changed. The reason for this is that the detector has an upper and a lower limit for the detection of opiates on a linear scale. Linear scales are used because these are the easiest to fit equations to for quantification purposes. Beyond the linear range of the response, the mathematical models become more difficult to apply.

A.6.4 The elution order of the barbiturates should remain constant, those with a more polar substitution at C-5 eluting before the more lipophilic compounds. The absolute retention times may not be identical. This may depend upon a number of factors, including the temperature of the column and of the solvent system (note that some systems have column ovens to maintain this part of the system at a constant temperature). The warmer the temperature, the lower the retention time.

Some barbiturate standards are notoriously difficult to obtain in a pure state. This is exemplified in the chromatogram shown (Fig. 10.4.) which exhibits a split peak for hexobarbitone. The quality of the separation will therefore also depend on the quality of the drugs being examined.

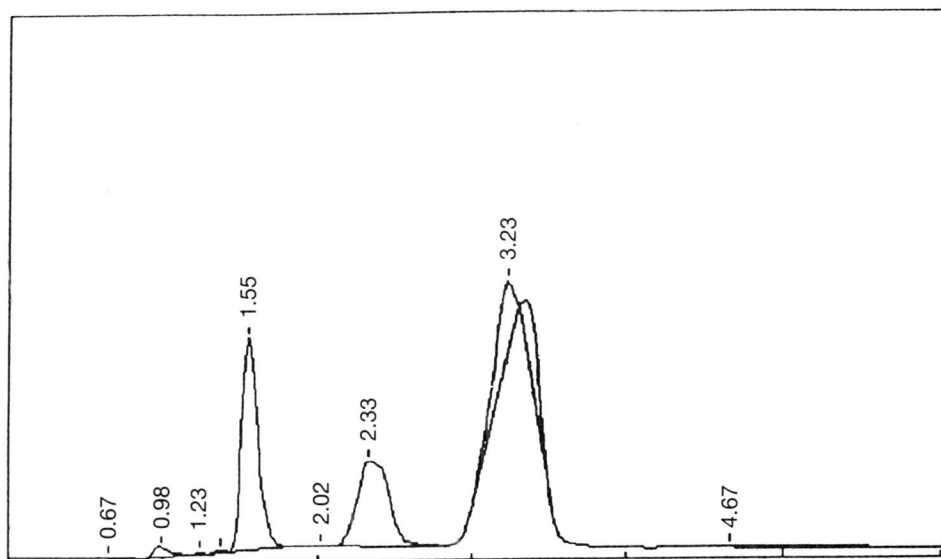

Fig. 10.4 Separation of barbiturate drugs using the HPLC system described,
(Elution order; barbitone, butobarbitone, hexobarbitone)

A.6.5 The retention data are provided in Table 10.4. The cannabidiol and cannabigerol both contain two hydroxide moieties and are the most polar. They are isomers of each other and are not resolved in this system. Cannabinol has only one hydroxide moiety, together with an unsaturated A ring. The electron cloud due to the pi electrons of this ring will contribute to the polarity of this molecule, but not to the same extent as the presence of another hydroxyl group. The A ring of the THC isomers is still more saturated than the cannabinol and hence these elute after this compound. Cannabichromene has two long side chains, making it the least polar of the compounds, hence it's high retention time.

Table 10.4 Retention times for cannabinoids using the HPLC system described

Cannabinoid	Retention Time (min)
Cannabidiol (CBD)	3.69
Cannabigerol (CBG)	3.69
Cannabinol (CBN)	5.55
Δ^9-Tetrahydrocannabinol	6.53
Δ^8-Tetrahydrocannabinol	6.53
Cannabichromene (CBC)	8.24

A.6.6 The effect of the acid will be to decrease the pH of the eluant (ie make it acid). All of the compounds to be analysed have acidic phenolic groups at the 1 position (and also at the 5 position of CBD and CBG). In aqueous solution, acidic groups such as this will dissociate resulting in charged species which interact with the stationary phase, resulting in tailing.

Addition of the dilute acid will force the drugs into the uncharged form. The non-dissociated nature of the molecules improves the chromatography because it reduces the effect of tailing due to interaction with residual silanol groups that may be present.

Chapter 7 Ultra violet spectroscopy of drugs of abuse

A.1 The general ground rules are that the absorption maxima of sample and control should match, as should the shoulders and troughs of spectra. In general, the general form of the overall spectrum should be the same. Additionally the isobestic point of the drug, if determined, should be the same.

Chapter 8   Infra red spectroscopy of drugs of abuse

A database should be relatively simple to build up.   It should then be possible to identify any drugs using the fingerprint region of the IR spectrum.

Chapter 9   Gas chromatography – mass spectroscopy of drugs of abuse

You should at this stage be able to determine whether the mass spectrum of the drug sample matches that of the control.   You should also be aware of the problems of weak samples with noisy baseline spectra, when using the NBS or any other GC/MS library.   Remember also that the library may not contain examples of derivatised drugs, and that there may have been variations in instrumental settings between the data that you have acquired, and for that stored in the library.